MIND MATTERS

evolution

MIND MATTERS

series editor: Judith Hughes

in the same series

Michael Bavidge, *Mad or Bad?*
Geoffrey Brown, *Minds, Brains and Machines*
Mark Corner, *Does God Exist?*
Grant Gillett, *Reasonable Care*
Ian Ground, *Art or Bunk?*
Mary Midgley, *Can't We Make Moral Judgements?*
Mark Thornton, *Do We Have Free Will?*
Robin Waterfield, *Before Eureka*

MIND MATTERS

evolution

ALEC PANCHEN

St. Martin's Press
New York

Scholarly and Reference Division,
St. Martin's Press, Inc., 175 Fifth Avenue,
New York, NY 10010

First published in the United States of America in 1993

Printed in Great Britain

ISBN 0-312-09697-6

Library of Congress Cataloging-in-Publication Data

Panchen, Alec L.
Evolution / A.L. Panchen.
p. cm.—(Mind matters)
Includes index.
ISBN 0-312-09697-6
1. Evolution (Biology) I. Title. II. Series.
QH366.2.P35 1993
575.01'6—dc20 93-18377
CIP

contents

foreword

'A philosophical problem has the form I don't know my way about,' said Wittgenstein. These problems are not the ones where we need information, but those where we are lost for lack of adequate signposts and landmarks. Finding our way – making sense out of the current confusion and becoming able to map things both for ourselves and for others – is doing successful philosophy. This is not quite what the lady meant who told me when I was seven that I ought to have more philosophy, because philosophy was eating your cabbage and not making a fuss about it. But that lady *was* right to suggest that there were some useful skills here.

Philosophising then, is not just a form of highbrow chess for postgraduate students; it is becoming conscious of the shape of our lives, and anybody may need to do it. Our culture contains an ancient tradition which is rich in helpful ways of doing so, and in Europe they study that tradition at school. Here, that study is at present being squeezed out even from university courses. But that cannot stop us doing it if we want to. This series contains excellent guide-books for people who do want to, guide-books which are clear, but which are not superficial surveys. They are themselves pieces of real philosophy, directed at specific problems which are likely to concern all of us. Read them.

MARY MIDGLEY

preface

In pondering the content of a useful preface, I began to wonder if I should start with an apology – whether it is necessary or not depends on my readers' (or potential readers') expectations of a book on evolution. If the word 'evolution' simply conjures up the proposition that humankind is descended from ape-like ancestors, then the apology is probably due. I shall say nothing about the details of human evolution, nor will I give a reconstruction of the evolutionary history of *Homo sapiens*, nor yet of all life. This is a book about the *theory* of evolution, a complex of interrelated scientific theories that explains a vast body of data and lower-level theories about living things and is supported by evidence from observation and experiment and tested by further observation and experiment.

If the evolution of living things, their change and diversification throughout the history of life, has occurred, then it is the job of evolutionary scientists to explain how that change and diversification came about. This book is about the why and how of evolution – *why* is it necessary to propose that evolution has occurred and is occurring, a historical theory, and *how* it is believed that evolution occurs, a theory of mechanism.

But like all science, evolutionary theory is a creation of the human mind – people observing, experimenting, thinking, suggesting explanations, but also disagreeing. I have tried therefore to give at least an impression of the history of the theory with some of the major figures involved in both past and present controversies as well as accepted advances. If (which I doubt) there is any branch of science where explanation and consensus are complete, that branch of sciences is dead, closed to further useful investigation. Evolutionary theory is not dead.

I am glad to acknowledge the help of friends and colleagues. I have benefited from discussions about evolution with several biological colleagues, particularly Professor Wallace Arthur, but also with members and associates of the former Department of Philosophy, University of Newcastle (including our editor) both before and after its philistine closure by the university. My zoological colleague Dr Henk Littlewood read and commented on most of the manuscript, which was typed by Yvonne Humble. The index was compiled by Harriet and Nicholas Luft. My thanks are due to all.

introduction

> Descended from apes! My dear, let us hope it is not so; but if it is, that it does not become generally known.
>
> (Attributed to the wife of the Bishop of Worcester, ca. 1860)

The theory of evolution enjoys a peculiar status in the culture of English-speaking peoples. As with methods of school-teaching and the way to run the economy, it is regarded as something on which everyone is entitled to an equal opinion, regardless of whether they are burdened with the necessary specialist scientific knowledge or not. In France, at least among professional biologists, a more robust view has been taken ever since Charles Darwin first proposed his theory. He committed the unpardonable sin of not being a Frenchman, so while most French biologists, at least by the end of the nineteenth century, accepted *that* evolution has occurred, they rejected Darwin's theory of Natural Selection, preferring the theory of their fellow countryman Lamarck to explain *how* evolution has occurred. For evolution implies not one theory but two*. The first theory is that the enormous diversity of animals, and plants and other organisms alive today is to be explained by cumulative change throughout the history of life on earth – a theory as to what has occurred. The second is a theory explaining how that change came about – a theory as to how it has occurred. These two theories were conflated in Darwin's original writings and the resulting confusion persists to the present day. Most of the first half of this book deals with the first theory, most of the second half with the second. But we must also ask the question: what is the theory of evolution for?

* Ernst Mayr, the Harvard biologist and doyen of students of evolution, suggests not two theories but five. These are distinguished in chapter 14, but are subsumed by my two.

Like any other scientific theory it must explain some puzzling body of data, something more than the statement that there are an awful lot of organisms. I shall also attempt to answer that question in the first part of the book. And there must be some evidence that evolution has occurred. This is reviewed in chapter 7.

My approach will be rather different from most books on evolution. The typical textbook approach, from secondary school biology texts to those for advanced undergraduates, is to begin with a detailed account of genetics, theories of heredity. The most generally accepted theory of mechanism, the how of evolution, is the mid-20th-century version of the theory of Natural Selection as originally proposed by Darwin and by Alfred Russel Wallace. That is then explained and illustrated by a few set-piece examples drawn from a remarkably limited pool. Then, at least in British school texts, there often follows something that strikes me as very odd. The theory of natural selection is contrasted with Lamarck's theory of mechanism (or more usually a parody of that theory) *just to demonstrate that Lamarck was wrong*! I can think of no other case where school children are expected to learn about a theory, first proposed nearly two hundred years ago, and are then instructed to reject it. Presumably the idea is to show the theory of natural selection in an even better light, which suggests some insecurity on the part of its advocates.

But there is something even odder about the standard textbook account, and that is a grievous error of omission. Having talked about natural selection (and Lamarckism) the remaining evolutionary topics are usually a section on 'evidence for evolution' followed by a systematic section demonstrating the diversity of organisms. This often takes the form of a reconstructed history of life on earth. The omission is any statement of the reason for evolutionary theory. What does the theory that evolution has occurred set out to explain? Why did Lamarck and Darwin and Wallace (and others) propose a theory of evolution in the first place? I shall begin chapter 1 with that question.

notes

Probably the best systematic account of modern evolutionary theory is in an undergraduate text **Douglas J. Futuyma**, *Evolutionary Biology* (2nd edn, Sinauer, 1986). **Roger Lewin**, *Human Evolution – An Illustrated Introduction* (2nd edn, Blackwell Scientific Publications, 1989) is readable and recommendable. But for the layperson to the professional it is difficult to find a wider perspective on evolution than the essays of **Stephen Jay Gould** in the American Museum magazine *Natural History*, published once a month. Some of these have been collected in a series of books, all but the last now in Penguin paperback:

Ever Since Darwin (1977), *The Panda's Thumb* (1980), *Hen's Teeth and Horse's Toes* (1983), *The Flamingo's Smile* (1985), *Bully for Brontosaurus* (1991) and *Eight Little Piggies* (1993).

1: what is the theory for?

natural classification

Chapters VII and VIII of Book IV ('Of operations subsidiary to induction') of John Stuart Mill's great work, *A System of Logic...* (first published in 1843), deal with classification. Both chapters deal almost exclusively with the classification of plants and animals, and Mill obviously sees something special about the classification of organisms, as distinct from that of either artifacts or other natural phenomena. Under a section heading *'Theory of natural groups'*, he writes as follows:

> The ends of scientific classification are best answered, when the objects are formed into groups respecting which a greater number of general propositions can be made, and those propositions more important, than could be made respecting any other groups into which the same things could be distributed...
>
> A classification thus formed is properly scientific or philosophical and is commonly called a Natural, in contradistinction to a Technical or Artificial, classification or arrangement.

In developing this theme Mill makes two points clear. The first is that there *is* such a thing as a 'Natural Classification' (his initial capitals) of organisms as distinct from an artificial one, the latter being a simple indexing and retrieval system such as the identification keys in books on natural history. Mill then expands on the properties of a natural classification. The most important of these concern consilience and prediction. The term 'consilience' is not Mill's but was coined by his arch-rival William Whewell (1794-1866), Master of Trinity College, Cambridge and noted polymath, who was 12 years older than Mill and disagreed with vigour and at length with most of his ideas on logic and the philosophy of science. They agreed, however, on the properties of a natural classification. One criterion in scientific inference to be applied to classification was expressed thus by Whewell:

> *The Consilience of Inductions* takes place when an Induction obtained from one class of facts, coincides with an Induction, obtained from another different class. This Consilience is a test of the truth of the Theory in which it occurs.
>
> (from *The Philosophy of the Inductive Sciences*, 1840)

In the days of Mill and Whewell animals and plants were classified using external features and the methods of comparative anatomy. But in the late 20th century evidence from embryology, molecular biology, biochemistry etc. can also be used. It is therefore possible to produce a series of classifications of the same set of organisms, each using a different technique. If these classifications turn out the same or closely similar, then the consensus classification has been justified by consilience. In the same spirit, if new characters (i.e. diagnostic features) are introduced into an existing classification, it should not alter the arrangement of organisms.

The criterion of prediction follows from Mill's definition of a natural classification. Let us imagine that on an expedition to Borneo, you the reader (being an expert on the classification of animals: i.e. an animal taxonomist) encounter a small warm-blooded, hairy quadruped. Without harming the animal you will be able to predict a number of observations that could be made by killing and dissecting it. It would prove to have a four-chambered heart from which came a left aorta taking oxygenated blood to the rest of the body. It would have a diaphragm, a muscular partition separating the chest from the abdomen. If the animal were a male you could even make a number of statements about the female of the species without even seeing her. She too would be hairy and warm-blooded etc., but would nourish her young with milk from her own mammary glands and (probably) give birth to live young. This is because all the features I have mentioned (except the last) are characteristics of mammals, the taxonomist's Class Mammalia. Strictly speaking, statements about the nature of the unseen female, and the internal anatomy of both sexes, are 'retrodictions' not predictions, and statements about the discovery of pre-existing phenomena are not of future events, but nevertheless they exemplify the predictive nature of Natural Classification.

Contrast this with the classification of library books. In most public libraries, including those in schools and institutions of higher education, books are arranged using a classification of the subject matter, unlike the situation in a small American university in which I once taught, where they opened a paperback section of the university bookshop, otherwise devoted almost exclusively to textbooks, and arranged the books in alphabetical order of title (there were an awful lot of books beginning with 'The'!). Thus if instead of looking for a living animal one wanted a book on the mammals of Borneo, it would be shelved under mammals, then vertebrates, then Zoology, then Biology, then pure (as distinct from applied) science. But the library classification would not have predictive properties, nor yet meet the criterion of consilience. The position in the library classification would tell one nothing about the size of the book (although it might have been separated for special shelving as a quarto or folio volume by a cursing librarian), nothing about the number of pages, price, colour and type of binding and (unless specifically programmed into the original classification) the language. So a library classification cannot be other than an example of Mill's Technical or Artificial classification. One can get no more information out of it than that which was programmed in.

the hierarchy

But a natural classification of organisms and an artificial classification of library books do have some features in common. The arrangement in both cases forms an *aggregational, inclusive, divergent, irregular hierarchy*. Most classifications, whether natural or artificial, take this form, simply because it represents the most convenient pattern for the retrieval of information; so most are created (to cite another jargon phrase) by a process of 'ordinally stratified hierarchical clustering'.

The pattern of classification is an *aggregational* hierarchy because it is built up by grouping the things to be classified, usually species in the case of organisms, into groups or clusters. The species or objects with which one starts then form the lowest rank of the classification, the initial clusters into which they are aggregated the next rank above. Thus in the library example above, the book on the mammals of Borneo, plus every other book in the library taken individually, occupies the lowest *rank* of the classification. Each book is then grouped with a small number of books on the same subject: in our example, books on mammals. All these primary groups or clusters throughout the library would then occupy the next rank. Beyond that the clusters of this second rank would themselves be clustered into a third and higher rank. In our example this would bring together all those clusters of books dealing with vertebrate animals, viz. the groups of books on mammals, the group on birds, the group on reptiles, the group on amphibia, the group on fish, plus a group on more general topics. The vertebrate book cluster so formed would then in its turn be clustered in a yet higher rank (Zoology) and so on. When the ranked clustering of all the books in the library had been completed the highest rank would be occupied by a single cluster consisting of the whole library (the books not the building!). The resulting ranked classification would be in the form of an *inclusive hierarchy*, inclusive because a cluster at any rank includes clusters at every rank below it.

A hierarchy need not be inclusive. In fact the technical term 'rank', used in the classification of organisms, suggests an example of an *exclusive* hierarchy. In the words of Ernst Mayr:

> Military ranks from private, corporal, sergeant, lieutenant, captain up to general are a typical example of an exclusive hierarchy. A lower rank is not a subdivision of a higher rank; thus lieutenants are not a subdivision of captains.

The hierarchy of classification is also *divergent*: no cluster at a lower rank belongs to more than one cluster at a higher rank. It could be the case that it was possible to classify a particular book in alternative widely-different high-rank clusters. Our book on the mammals of Borneo could be shelved under mammals, vertebrates...and so on, or, alternatively, under Borneo, South East Asia etc. within the section on geography. But in such cases librarians usually feel called upon to make a choice, at least if they only have one copy. So it is with the Natural Classification of organisms. If a species of animal or plant appears to belong to two different groups, one of those possible groupings is assumed to be a mistake. (We shall see later in this book that this view can be challenged.)

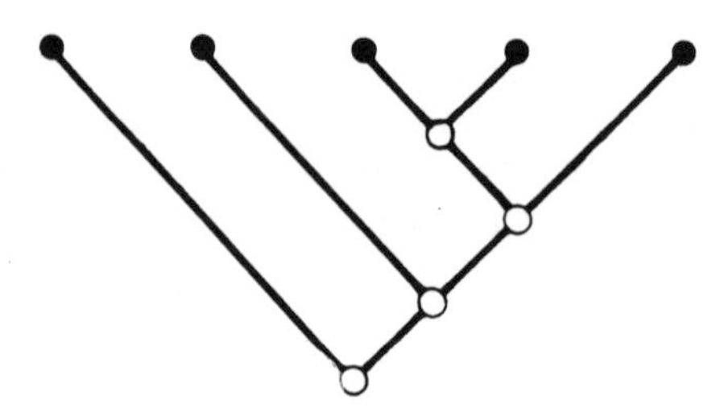

Fig. 1.1 A dendrogram – the branch points (open circles) and terminations of the branches (solid circles) are the 'nodes'.

As a result a classification (whether of library books or of animals) can be represented by the type of diagram known as a 'dendrogram' (Fig. 1.1). Each branching point on the dendrogram, (drawn as open circles) represents a group or cluster; clusters at the same rank are at the same level. The lowest rank (solid circles) is occupied by the individual books, or the animal species – those entities to be classified.

the *explanandum*

Perhaps, looking at the dendrogram, the patient reader will have guessed the answer to the question which heads this chapter: 'What is the theory for?' The dendrogram looks like a family tree, or at least the sort of family tree which shows all the descendants of a single patriarch. Such a family tree represents a pattern of relationship; one also talks of 'relationship' of animals and plants grouped together in a classification. But without any further assumption the word 'relationship' used to refer to the relative position of two species in a classification would be a metaphor; used literally it would imply evolution. Thus the theory that evolution has occurred is the theory that the apparent relationships of a natural classification are real relationships due to community of descent. If this is accepted the dendrogram representing a classification can be read in a different way. It can be read as *phylogeny*, a pattern of descent with evolution. All the branching points on the dendrogram can be seen as common ancestors, each the ancestor of all the actual species descendant from it. The dendrogram is now a 'tree' as that word is used by taxonomists. The all-inclusive single branching point or 'node' at the highest rank is the 'root' of the tree and, assuming evolution, represents the common ancestor of all the classified species.

One can say something about the nature of that common ancestor. It is probable that any feature shared by every species in the classification was a feature possessed by the ancestor of them all. The same principle applies to ancestors represented by nodes at lower ranks. Any given node will represent a hypothetical ancestor characterised by the features which all its descendant species have in common. It should be clear from this that any dendrogram of classification represents not just an inclusive hierarchy of groups, or *taxa* (singular 'taxon') as they are correctly termed, but a hierarchy of distinctive characters by which the members of any taxon are recognised. This is a point of very great importance and we will return to it (pp. 53-6).

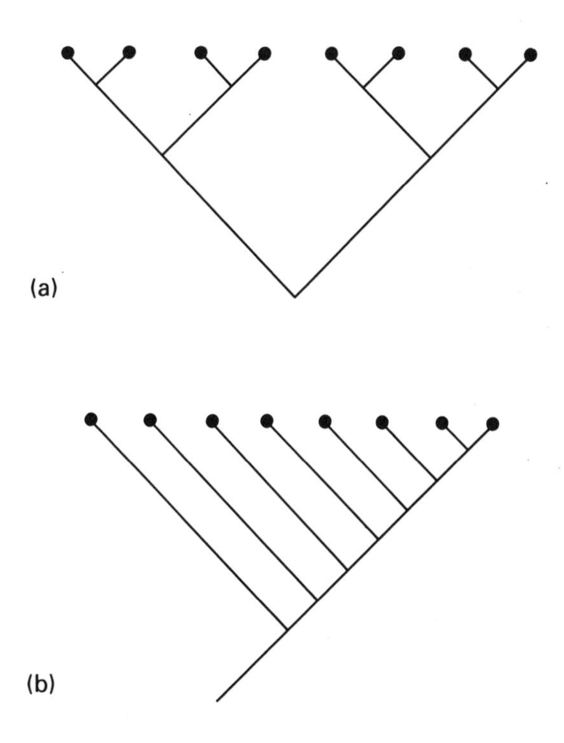

Fig. 1.2 The two regular extremes of dendrogram pattern (see text).

So far I have not given any account of one of the characteristics of a natural classification that I mentioned on p. 6. We have seen that the pattern of classification is represented by an *inclusive divergent hierarchy*, but I also said that it is *irregular.* One can envisage several types of regular dendrogram, and if the classification represented was of a small number of species the dendrogram might come out as regular by simple chance. But for a very large number of species irregularity is the norm. Two *regular* patterns are illustrated in Figure 1.2. The first (Fig. 1.2a) is a completely symmetrical dendrogram where every node has two descendent branches (dichotomous branching) and any sequence of branching has nodes at every rank. The dichotomous branching pattern may be an artifact resulting from the method of classification, but the regularity of ranks is taken to be real. At the other extreme is a completely asymmetrical pattern (again with dichotomous branching) (Fig. 1.2b). In this case each node has a different rank. This pattern is the conscious aim of some taxonomists but it cannot be sustained for a large number of species. As an aimed-for pattern it goes back at least to the ancient Greeks: we shall meet it again in chapter 3.

The fact that in most real cases the pattern is irregular leads to a very important conclusion. If all patterns of classification showed the same innate regularity, then one would be justified in looking for some principle of organisation which explained the apparent relationship of organisms. A good analogue is the Periodic Table of chemical elements. The Periodic Table was proposed by the Russian chemist Dmitri Mendeleef in 1869 who said that all the chemical elements could be arranged in a regular pattern based on their atomic masses: the modern arrangement is by atomic number, to which is related the number of planetary electrons which surround the nucleus of the atom. The elements in the table are then arranged in horizontal rows ('Periods'). Each row has 18 elements when complete. The next element after a run of 18 then begins the next row. Thus arranged each vertical column ('Group') contains elements with similar chemical properties (Fig. 1.3). The whole is explained by regularities in the number and arrangement of electrons,

Fig. 1.3 (see opposite) The Periodic Table of chemical elements with atomic numbers and atomic weights and standard initials for each element (from T.R. Dickson, *Introduction to Chemistry*, 6th edn, John Wiley and Sons, 1991 – by permission of the publisher).

Atomic number: 1, H, Atomic weight: 1.0079

Periods	IA 1	IIA 2	IIIB 3	IVB 4	VB 5	VIB 6	VIIB 7	VIII 8	VIII 9	VIII 10	IB 11	IIB 12	IIIA 13	IVA 14	VA 15	VIA 16	VIIA 17	Noble gases 18
1	1 H 1.0079																	2 He 4.00260
2	3 Li 6.941	4 Be 9.01218											5 B 10.81	6 C 12.011	7 N 14.0067	8 O 15.9994	9 F 18.99403	10 Ne 20.179
3	11 Na 22.98977	12 Mg 24.305											13 Al 26.98154	14 Si 28.0855	15 P 30.97376	16 S 32.06	17 Cl 35.453	18 Ar 39.948
4	19 K 39.0983	20 Ca 40.08	21 Sc 44.9559	22 Ti 47.88	23 V 50.9415	24 Cr 51.996	25 Mn 54.9380	26 Fe 55.847	27 Co 58.9332	28 Ni 58.70	29 Cu 63.546	30 Zn 65.38	31 Ga 69.72	32 Ge 72.59	33 As 74.9216	34 Se 78.96	35 Br 79.904	36 Kr 83.80
5	37 Rb 85.4678	38 Sr 87.62	39 Y 88.9059	40 Zr 91.22	41 Nb 92.9064	42 Mo 95.94	43 Tc (98)	44 Ru 101.07	45 Rh 102.9055	46 Pd 106.4	47 Ag 107.8682	48 Cd 112.41	49 In 114.82	50 Sn 118.69	51 Sb 121.75	52 Te 127.60	53 I 126.9045	54 Xe 131.29
6	55 Cs 132.9054	56 Ba 137.33	57 *La 138.9055	72 Hf 178.49	73 Ta 180.9479	74 W 183.85	75 Re 186.207	76 Os 190.2	77 Ir 192.22	78 Pt 195.09	79 Au 196.9665	80 Hg 200.59	81 Tl 204.383	82 Pb 207.2	83 Bi 208.9804	84 Po (209)	85 At (210)	86 Rn (222)
7	87 Fr (223)	88 Ra 226.0254	89 †Ac 227.0278	104 Unq (261)	105 Unp (262)	106 Unh (263)	107 Uns —		109 Une —									

58 Ce 140.12	59 Pr 140.9077	60 Nd 144.24	61 Pm (145)	62 Sm 150.36	63 Eu 151.96	64 Gd 157.25	65 Tb 158.9254	66 Dy 162.50	67 Ho 164.9304	68 Er 167.26	69 Tm 168.9342	70 Yb 173.04	71 Lu 174.967

90 Th 232.0381	91 Pa 231.0359	92 U 238.0289	93 Np 237.0482	94 Pu (244)	95 Am (243)	96 Cm (247)	97 Bk (247)	98 Cf (251)	99 Es (254)	100 Fm (257)	101 Md (258)	102 No (259)	103 Lr (260)

particularly those in the outermost orbit of each atom. The table is not a perfect rectangular grid, but all the anomolies are explicable and have been explained.

No such regularity exists, as far as we know, in the natural order of organisms. The irregular pattern of Natural Classification speaks therefore not of timeless organisation but of *contingency*. The properties of species, and of the individuals which make up a species are due to their history or the history of their forebears. Thus again we can say that the pattern of a natural classification represents a historical document culminating in the species being classified, the 'terminal taxa'. The first people to realise the significance of the irregularity of the pattern of natural classification were Charles Darwin and Alfred Russel Wallace, as we shall see in the next chapter. It is no coincidence that they were the first to propose both an acceptable theory that evolution had occurred *and* the basis of a valid theory of mechanism. But others, notably Lamarck, had already proposed theories of evolution. Lamarck, however, as we shall see in chapter 4, did propose some intrinsic organising principle.

In this chapter I have made the point that the *explanandum* of evolutionary theory, the data to be explained, is the natural order of organisms expressed in natural classifications, which are judged by the criteria of consilience and prediction. Such classifications are aggregational, inclusive, divergent, irregular hierarchies. The patterns of such classifications, dendrograms, can also be interpreted as 'trees' representing, in Darwin's phrase, 'descent with modification'. The technical term for such a pattern of descent is 'phylogeny' and it should be noticed that two components of evolutionary change are inherent within it. *Anagenesis*, evolutionary change with time, must be invoked, otherwise the hypothetical ancestor at the root of the tree would not differ from its ultimate descendants, the terminal taxa. But the terminal taxa differ from one another; divergence has occurred. It has occurred by *cladogenesis*, the splitting of one species into two or more, represented by the branching pattern of the tree.

notes

John Stuart Mill is still worth reading on classification (*A System of Logic...* Book IV, Chapters VII-VIII – the collected works edited by J.M. Robson, Toronto University Press, vol. 8, 1974).

The quotation from **Ernst Mayr** (on hierarchy) is from his magisterial *The Growth of Biological Thought* (Harvard University Press [1982] 205) that deals with the history of the study of 'Diversity, Evolution and Inheritance'.

Anagenesis: evolutionary change within a lineage.
Cladogenesis: branching of an evolutionary lineage, resulting from the splitting of one species into two or more (speciation).
Phylogeny: the pattern of evolutionary relationships of a group of taxa.
Taxon (pl. Taxa): a group of organisms (individuals or species) forming a unit at any rank in a classification.
Taxonomy: 'the theoretical study of classification...' [G.G. Simpson] (also sometimes used a mean [a] classification).

2: Darwin and Wallace

the irregular hierarchy

> [...]organised beings represent a tree *irregularly branched.* – Hence genera, as many terminal buds dying as new ones generated.
> (*Original emphasis*)

Darwin made this entry in the first of his private transmutation notebooks in the summer of 1837, eight or nine months after his return from the five-year round-the-world voyage of HMS *Beagle*, although the second sentence appears to have been added later. On later pages of the same notebook he doodled irregular trees (Fig. 2.1), and the sole illustration in *On the Origin of Species...* is a hypothetical tree-diagram which explicitly represents phylogeny. Darwin in the *Origin* (1859) is very careful to explain the intent of the diagram – such graphical representation was obviously regarded by him as something of a novelty. Not that his was the first graphical representation, either of the pattern of classification, or of phylogeny; Lamarck had produced both. Nor was Darwin the first to introduce the metaphor of the irregular tree, representing natural classification, to the public. He was preceded by Alfred Russel Wallace in a brilliant paper published by *The Annals and Magazine of Natural History* in 1855 and entitled 'On the law which has regulated the introduction of new species'. The law referred to reads '*Every species has come into existence coincident both in space and time with a pre-existing allied species*'. Although Wallace refers in the text to 'a pre-existing allied species' as the 'antitype' he meant 'ancestor', so that the whole includes a summary of the geographical and geological evidence for evolution (see ch. 7). But he also develops the tree metaphor:

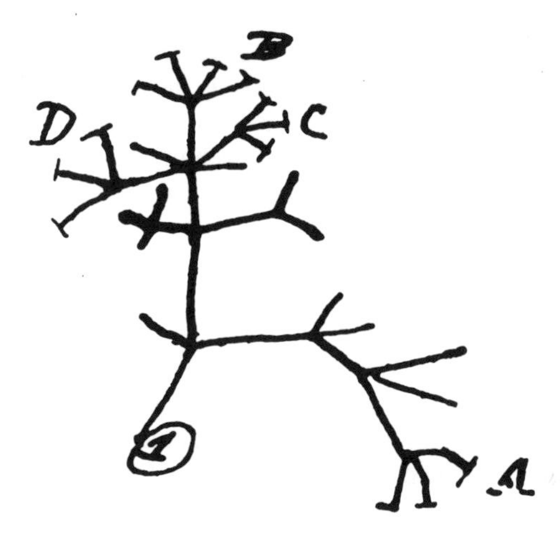

Fig. 2.1 An irregular phylogenetic tree from Darwin's first transmutation notebook (1837).

> ...in the actual state of nature it is almost impossible [to arrive at a true classification], the species being so numerous and the modifications of form and structure so varied, arising probably from the immense number of species which have served as antitypes for the existing species, and thus produced a complicated branching of the lines of affinity, as intricate as the twigs of a gnarled oak or the vascular system of the human body. Again, if we consider that we have only fragments of this vast system, the stem and main branches being represented by extinct species of which we have no knowledge, whilst a vast mass of limbs and boughs and minute twigs and scattered leaves is what we have to place in order...the whole difficulty of the true Natural System of classification becomes apparent to us.

Wallace's 1855 paper could be regarded as an (unintended) shot across Darwin's bows, warning him that another naturalist was also thinking about evolutionary theory and might pre-empt him in publication. It was so taken by Darwin's close colleagues Lyell and Hooker who warned him of the danger. A warning fulfilled in 1858 when Wallace sent Darwin the manuscript of a paper '*On the tendency of varieties to depart indefinitely from the original type*', which quite independently of the 'Big Book' that Darwin was writing marshalled the premises and reached the conclusions of Darwin's theory of Natural Selection as the mechanism for evolution. Darwin's and Wallace's backgrounds were very different, but they shared with many other 19th-century naturalists a training in their discipline resulting from long and arduous expeditions abroad.

Darwin

Charles Robert Darwin was born on 12 February 1809 at Shrewsbury, the son and grandson of prosperous medical practitioners. His famous grandfather Erasmus Darwin had speculated on the common ancestry of all organisms and sketched a Lamarckian mechanism in a medical treatise, *Zoonomia.* Charles was at school in Shrewsbury and, following family tradition, was sent to read medicine at Edinburgh University in 1825. Bored by his strictly medical studies and appalled by witnessing operations performed without anaesthetics, he withdrew in 1827 and was sent to Christ's College, Cambridge, where he read for an ordinary BA degree, with the intention of entering the church, and graduated in 1831. But while not distinguished in his studies as an undergraduate, as an amateur naturalist he attracted the attention of the Rev. Adam Sedgwick, professor of geology, who gave him a practical grounding in field geology, and of the Rev. J.S. Henslow, professor of botany, who was later to recommend him for his role on the *Beagle*. He embarked on the *Beagle* on 27th December 1831 as naturalist-companion to the captain, Robert Fitzroy. The *Beagle* was a surveying vessel, particularly charged with charting the coasts of South America. This task occupied her until September 1835, during which time Darwin made a number of overland journeys. She then sailed to New Zealand and Australia, via the Galapagos Islands, and returned home by way of South Africa and South America, arriving at Falmouth in October 1836. Darwin opened his first 'species notebook' in July 1837, but his published work between then and 1859, when the *Origin* appeared, was mostly geological, some of it based on the *Beagle* voyage. In addition he contributed to the scientific accounts of the natural history of the journey, and wrote a memoir of his adventures. Notably, however, he spent the period between 1846 and 1854 preparing four large monographs on the classification of fossil and living barnacles.

Those monographs are still of scientific value today and his book on coral reefs and atolls, first published in 1842, is the base from which modern theories of their formation were developed. Soon after he published books on volcanic islands and on the geology of South America.

But during this period he was refining his 'species theory', first written as a private pencil sketch in 1842 and then enlarged into a more extended, but still secret essay in 1844. The barnacle work completed, Darwin started on his 'big book' (part of which has been reconstructed and published fairly recently). This project was forestalled, in one of the legendary incidents in the history of science, when he received Wallace's manuscript '*on the tendency of varieties...*' referred to above, together with a letter from Wallace asking Darwin to consider it for publication. In something of a panic, Darwin handed the whole matter over to his two friends and colleagues, Charles Lyell, the geologist, and Joseph Dalton Hooker, the botanist. Without waiting for Wallace's approval they arranged for Darwin's 1844 essay, Wallace's paper and an abstract letter Darwin had written to the Harvard botanist Asa Gray in 1857 about his theory (to establish Darwin's priority) to be read before the Linnean Society of London on 1st July 1858. In the following months Darwin prepared an 'abstract' of his theory which was published on the 24th November 1859 as *On the Origin of Species by means of Natural Selection, or the preservation of favoured races in the struggle for life*. Popular mythology has it that all the copies were sold out in a single day. In a sense this is true, but it does not mean that an eager Victorian public snapped up all those copies for sale but merely that all were taken up by the booksellers. This was just as well in that John Murray (who had also published Darwin's *Journal of Researches...* i.e. 'The Voyage of the *Beagle*') published the *Origin* at author's risk. The *Origin* went through six editions plus a final corrected reprinting published in 1876.

By the time of the publication of the *Origin*, Darwin had effectively confined himself to his home in Kent, venturing out mostly for ineffective hydropathic cures. Cossetted by a loving family, but plagued by debilitating gut disorders and other symptoms which were exacerbated by social excitement, he avoided any face-to-face controversy which his theory would have generated at scientific meetings etc. Despite this he maintained an enormous correspondence, acted as a benevolent squire in the village of Downe and published ten more books and some hundred scientific papers. His subsequent books were written as a conscious research programme, developed out of the theories presented in the *Origin*. Notably *The Variation of Animals and Plants under Domestication* (2 vols: 1868) was originally intended as a part of the big book and, strangely, was first published in parts commencing in 1867 in Russian translation as *On the Origin of Species. Section 1*. Although the *Origin* did not deal with human evolution, the public quickly grasped the implications of Darwin's theory, and in 1871 he produced *The Descent of Man, and Selection in Relation to Sex*: two subjects in one (double) volume, followed in 1872 by *The Expression of the Emotions in Man and Animals*. Apart from a long preface to a biography of his grandfather and his book on the action of earthworms, published in the year before his death, the remaining books were on aspects of botany. Charles Darwin died on 19th April 1882 and, rather against the wishes of his family, was buried with due pomp in Westminster Abbey.

Wallace

Alfred Russel Wallace was born in Usk, Monmouthshire on the 8th January 1823, the seventh of nine children of an improvident but affectionate father. His schooling in Hertford, whence the family had moved in 1828 lasted only until he was thirteen, after which he was largely self-educated. For most of the time between then and 1848, when he embarked on his first great expedition, he was employed as a surveyor, initially as an apprentice to his brother William, who died in 1845. Wallace also had a spell as a schoolmaster in Leicester and in the building trade with his brother John.

While in Leicester in 1844 Wallace became friendly with a fellow amateur naturalist Henry Walter Bates; this was to lead to their joint plan to go to the Amazon region to study the rain forest and solve 'the species problem'. They left for Brazil in 1848 where they were joined by Wallace's brother Herbert (who died of yellow fever in Amazonia after two years) and Dr Spruce, a botanist, in 1849. Wallace and Bates separated to explore different regions in 1849. Wallace left Brazil in 1852: Bates remained until 1859. Both published very readable accounts of their adventures.

Despite the tragedy of his brother's death, and a calamity when his returning ship caught fire at sea and was lost with most of his personal collection, Wallace again set sail for the tropics in 1854, this time for the 'Malay Archipelago' (the title of his account of his travels published in 1869) where he stayed until he returned to London in 1862. He supported himself in the Far East as he and Bates had done in Amazonia, by selling natural history specimens through an agent. Thus both of Wallace's important papers on evolution, the 'Sarawak essay' of 1855, and his proposal of natural selection in 1858, written on one of the Molucca Isles, were completed when abroad.

Like Darwin, Wallace wrote a number of books on his return, but the titles seem oddly assorted. He always gave priority for their great theory to Darwin, but published two books on the subject, *Contributions to the Theory of Natural Selection* (1870), mostly reprinted essays, and *Darwinism...* (1889). After his return Wallace's principal scientific contribution was in biogeography (see ch. 7) including a magisterial two-volume work *The Geographical Distribution of Animals* (1876). But unlike Darwin, Wallace was a great supporter of causes, from the worthy to the silly. Thus he published on socialism and land nationalisation, on general social conditions, on spiritualism (after, to Darwin's distress, he ceased to accept the natural origin of the human mind) and a series of diatribes against vaccination. Formal honours came to Wallace late in life: he was elected to the Royal Society at the age of 69. He did not die until his 91st year, in November 1913.

the *explanans*

I began this chapter to emphasise that Darwin's and Wallace's proposals of a successful evolutionary theory (both phylogeny and mechanism) were due in part to their shared realisation of the irregularity and thus contingent nature of the pattern of natural classification. Both of course did much more than this, and most historical accounts would accord it little importance if mentioning it at all. Darwin's and Wallace's other achievements were to present compelling evidence that evolution had occurred and to propose a mechanism: both these are dealt with later in this book. But there is more that needs to be said about classification and the characters on which it is based, in order to show the intimate and important relationship between evolutionary theory and taxonomy. That will be the subject of our next four chapters. Meanwhile, to emphasise the central position of the tree metaphor in Darwin's thinking (and to hint at the possibility that he copied it from Wallace) I quote from the long last paragraph of Chapter 4 of the *Origin*.

> The affinities of all the beings of the same class have sometimes been represented by a great tree. I believe this simile largely speaks the truth. The green and budding twigs may represent existing species; and those produced during each former year may represent the long succession of extinct species. At each period of growth all the growing twigs have tried to branch out on all sides, and to overtop and kill the surrounding twigs and branches, in the same manner as species and groups of species have tried to overmaster other species in the great battle for life. The limbs divided into great branches, and these into lesser and lesser branches, were themselves once, when the tree was small, budding twigs; and this connexion of the former and present buds by ramifying branches may well represent the classification of all extinct and living species in groups subordinate to groups. Of the many twigs which flourished when the tree was a mere bush, only two or three, now grown into great branches, yet survive and bear all the other branches; so with the species which lived during long-past geological periods, very few now have living and modified descendants. From the first growth of the tree, many a limb and branch has decayed and dropped off; and these lost branches of various sizes may represent those whole orders, families, and genera which have now no living representatives, and which are known to us only from having been found in a fossil state....

Natural Classification was the *explanandum* of Darwin's theory: the *explanans*, a part of the theory itself, was phylogeny.

notes

The best way, even for the layperson, to come to terms with the work of **Darwin** and **Wallace**, is to read both in the original. *On the Origin of Species...* is probably more accessible than any other important primary work of science. The first edition (1859) is strongly recommended: Darwin was subsequently assailed by (mostly unnecessary) doubts which weakened the impact of later editions. It has been reprinted many times. The submission of

Darwin's and Wallace's theories to the Linnean Society (1858) was reprinted as *Evolution by Natural Selection* in 1958 (ed. G.R. de Beer, Cambridge University Press). Wallace's 1855 paper 'On the law which has regulated the introduction of new species' is in *The Annals and Magazine of Natural History* (n.s.) 16: 184-96.

The 'Darwin industry' is notorious for the obsessive scrutiny by innumerable authors of every aspect of his life and work. I will recommend only **D. Ospovat,** *The development of Darwin's Theory: Natural History, Natural Theology and Natural Selection 1838-1859* (Cambridge University Press, 1981). **H. Lewis McKinney,** *Wallace and Natural Selection* (Yale University Press, 1972) deals with the life and work up to the proposal of the theory.

3: a historical–philosophical interlude

the *scala naturae*

> ...all the orders of natural beings form but a single chain, in which the various classes, like so many rings, are so closely linked one to another that it is impossible for the senses or the imagination to determine precisely the point at which one ends and the next begins – all the species which, so to say, lie near to or upon the borderlands being equivocal, and endowed with characters which might equally well be assigned to either of the neighbouring species.
>
> (*Leibniz*, in a letter quoted by A.O. Lovejoy, *The Great Chain of Being*)

> If an evolutionary continuum existed, as the evolution model should predict, there would be no gaps, and thus it would be impossible to demark specific categories of life. Classification requires not only similarities, but differences and gaps as well, and these are much more amenable to the creation model.
>
> (H.M. Morris, *Scientific Creationism*, 1974)

I begin this chapter with a confession! In chapter 1, I drew attention to John Stuart Mill's characterisation of a natural classification with its properties of maximum consilience and predictive value. I also quoted from Chapter VII of Book 4 of his *System of Logic* as I, and others interested in classification, have done on previous occasions. It was only recently, however, that I went on to read his Chapter VIII, the second one on classification. When I did so I received a considerable shock. He appeared, in 1843, to be advocating an ancient method of ordering organisms, which was at odds with everything in the previous chapter. Chapter VIII ('Of classification by series') begins:

> Thus far, we have considered the principles of scientific classification so far only as relates to the formation of natural groups; and at this point most of those who have attempted a theory of natural arrangement, including, among the rest, Dr Whewell [!], have stopped. There remains, however, another, and a not less important portion of the theory, which has not yet, as far as I am aware, been systematically treated of by any writer except M Comte [the inventor of 'positivist' philosophy]. This is, the arrangement of the natural groups into a natural series.

Whewell responded to the implicit criticism saying that he 'stopped short of, or rather passed by, the doctrine of a series of organised beings', because he 'thought it bad and narrow philosophy' to claim that the natural order was 'a mere linear progression in nature, which would place each genus in contact only with the preceding and succeeding ones'. Whewell was in fact rejecting the ancient doctrine of the *scala naturae*, or Great Chain of Being.

Mill returned to the fray in a footnote in the 3rd edition of the *System of Logic* (1851). Whewell had misread his intention:

> Now the series treated of in the text agrees with this linear progression in nothing whatever but in being a progression. It would surely be possible to arrange all *places* (for example) in order of their distance from the North Pole, though there would be not merely a plurality, but a whole circle of places at every single gradation in the scale.

What Mill was advocating was not a linear series of species (or genera, the next ranking category of taxa) but a linear series of higher ranking groups.

I revive this exchange between two dead philosophers, because it illustrates two forms of a concept which has dogged the search for the natural order for at least two thousand, three hundred years and still will not die. The first, the *scala naturae*, goes back to the ancient Greek Philosophers. The second, Mill's concept of a linear series of major groups, took over during the nineteenth century, especially when biologists started to claim that evolution (i.e. phylogeny) was the basis of their classifications (see ch. 5).

The concept of the *scala naturae* is usually attributed to Aristotle (384-322 BC) and was developed by the 'Neoplatonists' in the first few centuries of the Christian era. The idea is that all living things, or at least either all species or all genera, can be arranged in a linear series of increasing perfection, with mankind at the top and the simplest organisms at the bottom. There were variations: sometimes the series was extended further 'downwards' to inanimate objects or further upwards to the (exclusive) hierarchy of angels, archangels etc.; or again a philosopher or naturalist might postulate separate *scalae* for animals and plants.

But almost always the concept of the *scala naturae* was accompanied by another, that of *plenitude*, attributed by some to Plato. This was the doctrine that if creation is perfect, it should contain no gaps, so the ascending series of the *scala* was unbroken: each species or genus merged into those on either side of it.

The *scala naturae* was very popular, both with philosophers and naturalists, in the 17th and 18th centuries (see our epigraph from Liebniz – and from Morris, whose mode of thought is that of the 17th century). Notable among the latter was the Swiss naturalist Charles Bonnet. His *scala* was expressed merely in words, but has been produced as a diagram subsequently (Fig. 3.1). From a 20th-century viewpoint acceptance of the *scala naturae* might seem to pose problems for the taxonomist who wished to produce a classification which was a hierarchical clustering, particularly if he wished to claim that such a classification was natural. Yet the formalisation and codification of methods of classification dates from the middle of the 18th century when the popularity of the *scala* was at its height. It was due largely to the great Swedish naturalist Linnaeus and a positive army of collectors who worked for him. Linnaeus seems to have accepted the *scala naturae* for animals, though plants gave him great trouble. At the same time he proposed a classification of animals, and in part of plants, which forms the basis of those used today. The formal 'starting point' for valid animal classification is 1758, representing the publication of the 10th edition of his *Systema Naturae*, vol. 1, which contains his fully developed and, as he would have claimed, natural classification of animals.

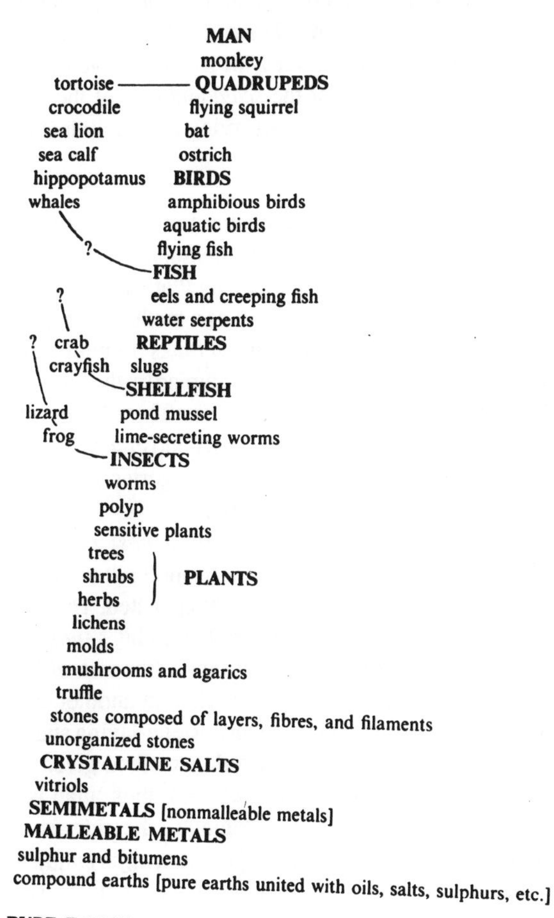

Fig. 3.1 *Idée d'une échelle des êtres naturelles* – drawn as a diagram from Bonnet's 1764 list by P.C. Ritterbush in *Overtures to Biology*, Yale University Press, 1964 – by permission of the publishers.

But Linnaeus' claim for the naturalness of his classification depended not so much on the naturalness of the result, but on the correctness of the method. Hence the fact that he and others, including perhaps John Stuart Mill a hundred years later, could subscribe to the concepts of the *scala* and also of the natural hierarchical classification without worrying about any inconsistency.

logical division

Despite his reputation as an innovator, Linnaeus' system of classification was taken over virtually unmodified from the ancient Greeks. It was originally elaborated by Plato and Aristotle not as a means of classifying organisms but as a method of gaining knowledge. As we have seen the two lowest ranks normally encountered in a classification of organisms are occupied by the '*categories*' (the correct technical term) *genus* and *species*. These Latin words correspond to the ancient Greek terms *genos* and *eidos* respectively and this latter pair are the fundamental categories in the method of *logical division* of Plato and Aristotle.

The original use of the term *genus* (or *genos*) was not to name a category at a particular rank in classification but to label a class of objects whose properties were to be investigated. Thus Aristotle in one case took the whole animal kingdom as his *genos* and divided them into animals with or without blood (the two *eide*). He also divided the kingdom in other ways. Then in another case he took snakes as his *genos* and divided them into egg-laying and viviparous forms. The usual example of Aristotle's method given in books is his treatment of man (i.e. *Homo sapiens*): thus 'man is a rational animal'. In this case the *genos* is 'animal', the *eidos* is 'man', and 'rational' is known as the *diafora* (in Latin, *differentia*). A 'definition' of man then consists of a statement of the *genos* and the *diafora*. Other attributes of man were known as properties – Aristotle's (dubious) example was 'capable of learning grammar'. As a system of gaining knowledge, the whole was rather spoiled by another type of attribute, 'accidents'. These either did not describe all the members of the *eidos*, or did not apply all the time – 'in a sitting position' was one of Aristotle's examples. Because there was no certain, i.e. logical, way of distinguishing properties from accidents by investigating a sample of the *eidos*, the system failed as a means of acquiring new knowledge. But it contains the essence of some important features of taxonomy and was taken over by Linnaeus (and in part by some of his predecessors). Thus by treating genus (*genos*) and species (*eidos*) as fixed taxonomic categories, all organisms can be named. Each animal or plant species is given a generic name (e.g. *Homo*) and a specific name (*sapiens* – i.e. 'wise' or 'rational'!). In formal nomenclature, the name of the original describer is then appended – *Homo sapiens* Linnaeus (usually abbreviated 'L.' or 'Linn.'). Linnaeus also took over the Aristotelian idea of the *differentia*. Each species was distinguished from the others in the same genus, by its own *differentia*, often known as its diagnosis, and consisting (in Linnaeus' *Systema Naturae*) of a description in a sort of Latin shorthand of the characteristic features of the species. Genera and higher ranking categories were similarly defined.

So in Linnaeus' system of classification, we have a standardised system for nomenclature in the binomial system (generic plus specific name). We also have standardised short-hand description of any taxon, both at the level of species and that of genus ('*Homo*' is a taxon of the category or at the rank of genus, '*sapiens*' is a taxon of the category or at the rank of species). But Linnaeus also gave his classifications the formal structure of an inclusive, divergent hierarchy, and that hierarchical structure was again taken straight from the traditions of Greek philosophy.

In his late work, the *Sophist*, Plato sets out a hierarchical classification in which the occupation of angling, is classified among the human arts by a series of dichotomous divisions (Fig. 3.2). I would particularly like the reader to notice that the pattern of this classification, the first of its kind that has come down to us, is the same as the most asymmetrical cladogram which I illustrated in Figure 1.2. That pattern will recur in this book. What Plato did not do was to label each rank in his classification with a category name. Later, however, a classification with standard categories was produced in the early Christian 'Neoplatonist' tradition of that group of philosophers who tried to reconcile and rationalise the works of Plato and Aristotle. It is known after its reputed author Porphyry (ca. AD 232-304) as the 'Tree of Porphyry' (Fig. 3.3). The pattern is asymmetrical as in Plato's classification of angling as one of the arts, but the category at each rank is named thus: *summum genus*, *subaltern genus* (at three ranks), *Infima species* and *Individua*, where each *individuum* is a member of the species 'man'. Subaltern genera could presumably be expanded or

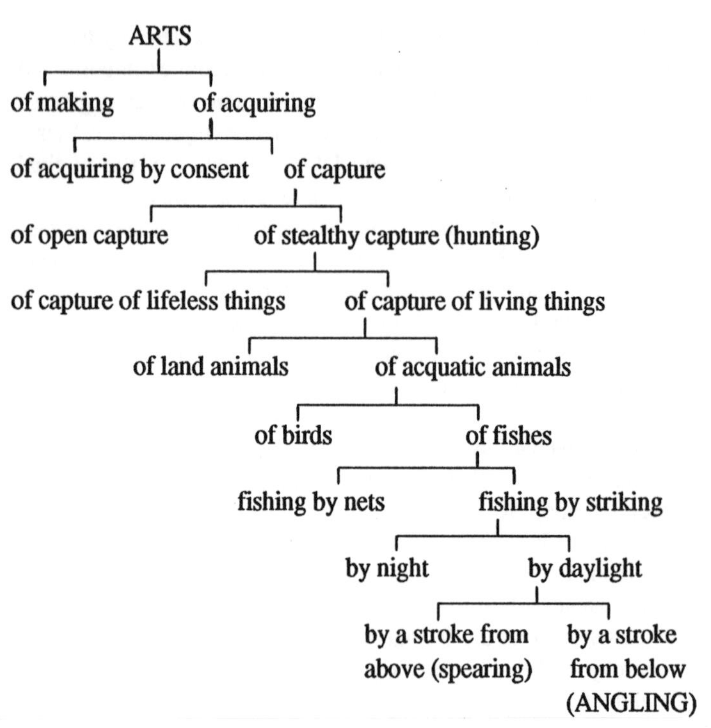

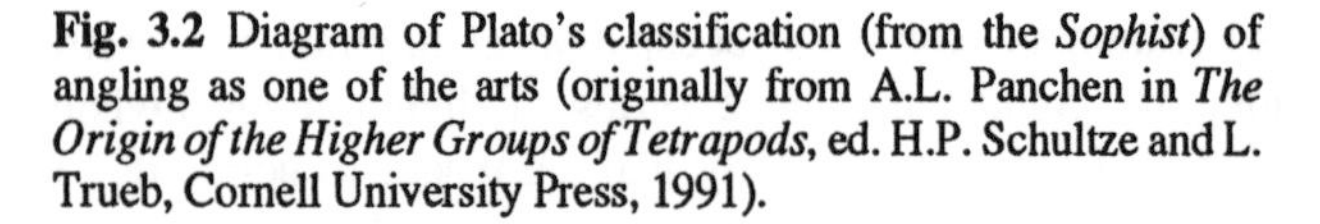

Fig. 3.2 Diagram of Plato's classification (from the *Sophist*) of angling as one of the arts (originally from A.L. Panchen in *The Origin of the Higher Groups of Tetrapods*, ed. H.P. Schultze and L. Trueb, Cornell University Press, 1991).

contracted in number to suit any individual classification. When Linnaeus established his hierarchy he took over Porphyry's system in its entirety, except that he adopted just two levels of subaltern genera, which he named '*genus intermedium*' and '*genus proximum*'. In his *Systema Naturae* these categories are then parallelled with a series of biological category names introduced *en bloc* by Linnaeus and still used today:

Genus summum	*genus intermedium*	*genus proximum*	*species*	*individuum*
Classis	*Ordo*	*Genus*	*Species*	*Varietas*

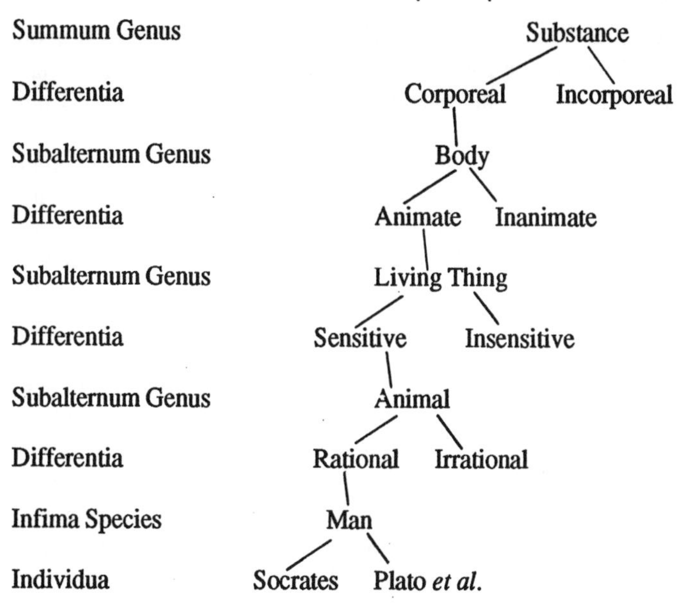

Fig. 3.3 The 'Tree of Porphyry' with categories and differentia (originally from A.L. Panchen, 1991, *loc. cit.*).

Linnaeus' classes of animals were *Mammalia*, *Aves* (birds), *Amphibia* (including reptiles), *Pisces* (fishes), *Insecta*, and *Vermes* (literally worms, but including all the remaining invertebrates). Most classification today requires more than four ranks, and thus categories, to species level. Kingdom ('*Regnum Animale*') predated Linnaeus but Phylum, between Kingdom and Class, and Family, between Order and Genus, were introduced in the 19th century, as well as categories for intermediate ranks designated by 'super-' (e.g. Superorder), 'sub-' (Suborder), 'infra-' (Infraorder), and latterly some less regularly used categories such as 'cohort'.

taxonomy so far

The aim of this chapter was to show that by the end of the 18th century, following Linnaeus' work, there was an agreed method of presenting a classification with a standard hierarchy and a named category for each rank, a standard system of nomenclature, and a method of characterising every taxon by a diagnosis. Linnaeus believed that by following the correct procedure (known to him as logical division *per genus et differentiam*) a 'natural' classification would result. The pattern of such a classification was a divergent hierarchy even if the natural order was perceived to be a *scala naturae*.

In chapter 2 I set out to show that, given that the natural order (and its representation as the pattern of classification) is an inclusive, divergent, irregular hierarchy, the irregular and thus contingent nature of the pattern was emphasised both by Wallace and by Darwin. In the classification of Linnaeus, at least of animals, the pattern is that of a divergent hierarchy, but it is not certain whether he saw this as the natural order, in the sense of a real phenomenon in nature as distinct from a human construct. Furthermore, we do know that Linnaeus believed that the only way in which new species *may* have arisen was by hybridisation, which would have destroyed the overall divergent pattern. Between the time of Linnaeus and that of Darwin, a number of distinguished naturalists, particularly in France, contributed ideas which affirmed the natural status of the divergent hierarchy. They are the subject of our next chapter.

notes

The classic essay on the *scala naturae* is **A.O. Lovejoy**, *The Great Chain of Being* (Harvard University Press, 1936). **David Oldroyd's** *The Arch of Knowledge* (Methuen, 1986) is an excellent account of 'the history of the philosophy and methodology of science' that describes the epistemology of Plato and Aristotle and the subsequent development of the ideas of the *scala* and plenitude. Linnaeus' use of logical division is described by **A.J. Cain**, 'Logic and memory in Linnaeus' system of taxonomy' (*Proceedings of the Linnean Society* 169 [1958] 144-63).

4: the other French revolution

the French connection

The 'Reign of Terror' in the French revolution reached its climax between the early days of October 1793 and the end of that year. Less than two months before, the organisation of the *Museum d'Histoire Naturelle* in Paris had been reconstituted from that of its predecessor the *Jardin du Roi* (or, since 1790, *Jardin des Plantes*). Among those who had taken part in planning the new institution was the Chevalier de Lamarck, known up to that time as a botanist and author of the *Flore françoise*. It was not just to be a museum and botanic garden, but also a foundation for research and teaching, with twelve professorial chairs in different branches of natural history. Lamarck in his fiftieth year was switched from botany to be the professor of 'insects, worms and microscopic animals'. Vertebrate animals were at first the sole charge of the twenty-one-year-old Etienne Geoffroy Saint-Hilaire but in the next year he was confined to mammals and birds, with Bernard de Lacépède as professor of reptiles and fishes. Late in 1794 a young naturalist, George Cuvier, then a private tutor in Normandy, but born a citizen of the Duchy of Württemberg, was recommended to Geoffroy, who invited him to Paris, where he arrived in 1795. After various teaching jobs Cuvier was appointed professor of comparative anatomy at the Museum in 1802.

The three great naturalists, Lamarck, Geoffroy St.- Hilaire and Cuvier, each contributed important data and ideas in the transition from the Aristotelian view of classification of Linnaeus to the interpretation of the pattern of classification as phylogeny by Darwin and Wallace. But while the three were colleagues at the Museum of Natural History for twenty-seven years, theirs was not a happy association. The ageing Lamarck increasingly became a recluse and was despised by Cuvier who differed from him on almost every point of theory. For the last ten years until his death in 1829 Lamarck was blind and most of his great *Histoire naturelle des Animaux sans Vertèbres*, published between 1815 and 1822 was copied out by his daughter. Geoffroy and Cuvier began as close colleagues, friends and collaborators, but their views on comparative anatomy diverged, and in 1830 their differences burst out in a famous debate, a spectacular public row that went on until Cuvier's death in 1832.

Lamarck and evolution

Both Cuvier and Geoffroy rejected the concept of the *scala naturae* and so, at least in its simple form, did Lamarck at the end of his career. But Lamarck's first theory of evolution, probably arrived at in the year 1800 and presented in his lectures in 1801, was based on the *scala*. In the late 18th century there was a great fascination with both electricity, generated by Leyden jar, and magnetism. Heat, electricity and magnetism were each thought to be the manifestation of 'subtle fluids'. That related to heat was 'caloric'; a subtle fluid was also said to be responsible for 'animal magnetism' as proclaimed by Mesmer and his 'mesmerism'. Towards the end of the century, it was one of Lamarck's aims, much scorned by his French scientific colleagues, to unite all science in one great system. It is perhaps not surprising, therefore, that he proposed 'subtle and ponderable fluids' as the driving force of evolution. Thus in his original evolutionary theory, the '*marche de la nature*', the *scala naturae* was (as it were) turned into an escalator: thus in 1802 –

> Ascend from the simplest to the most complex; leave from the simplest animalicule and go up along the scale to the animal richest in organisation and facilities...then you will have hold of the true thread that ties together all of nature's productions, you will have a just idea of her *marche* and you will be convinced that the simplest of her living productions have successively given rise to all the others.

But where did the 'simplest animalicule' come from? Lamarck had the answer:

> ...once the difficult step [of accepting spontaneous generation]

> is made, no important obstacle stands in the way of our being able to recognise the origin and order of the different productions of nature.

Perhaps the most remarkable feature of Lamarck's original theory was that it was a steady-state theory of evolution. Simple animalicules were arising all the time by spontaneous generation. Driven by subtle fluids, their descendants evolved along the pre-ordained track of the *scala naturae*, with organisms dying and decaying to provide the raw material for more spontaneous generation, until the stage of mankind was reached. Once the human stage had been reached the species present in the world would remain the same, despite the fact that each individual line of descent was evolving. Lamarck did not accept extinction as a general phenomenon, although he allowed that it might occur in newly-generated '*Monas*' or by human agency. But while Lamarck was developing his theory of the 'escalator naturae' from 1800, he also, almost as a separate exercise, developed the theory of evolutionary mechanism for which he has become famous – 'the inheritance of acquired characters'. I will describe this in chapter 8, but meanwhile we must look at Lamarck's contribution to the perception of the natural order as phylogeny.

The principal presentation of his views on evolution and 'biology' (a word coined by Lamarck) was in the *Philosophie Zoologique* of 1809, but this contains some internal contradictions. On the one hand Lamarck distinguishes between two sorts of classification, '*distribution gènèrale*', which is placing organisms (particularly animals) in their correct position on the *scala naturae* – seen as a search for the natural order in contrast to '*classification*' into a hierarchy, seen as artificial; but on the other hand he concludes that:

> It is there shown [in a diagram] that in my opinion the animal scale begins by at least two separate branches, and that as it proceeds it appears to terminate in several twigs in certain places.

Later in his *Natural History of the Invertebrates*, Lamarck had a somewhat more complex diagram (Fig. 4.1). The branching pattern again had two separate roots and represented the pattern both of classification and phylogeny, but animals were also divided into

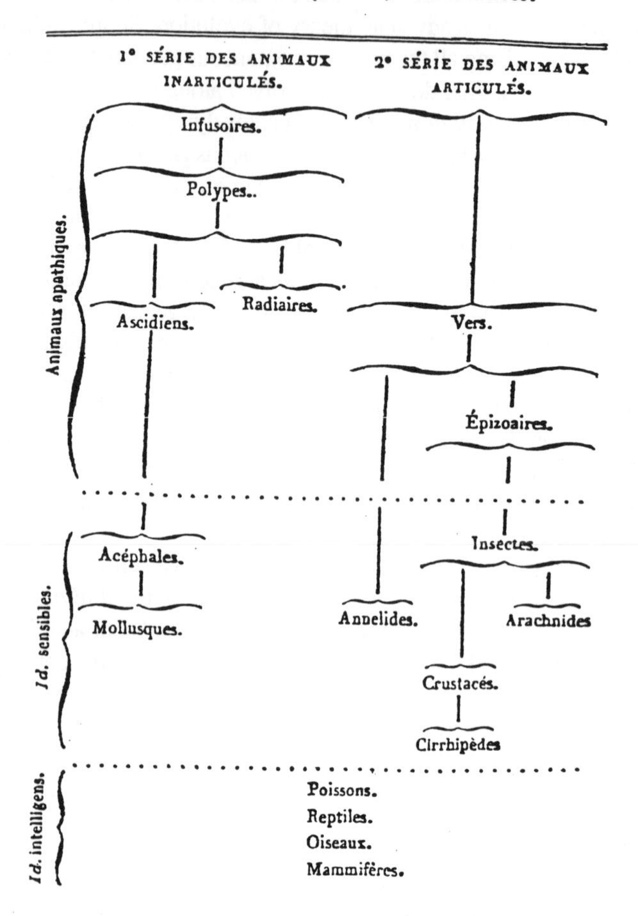

Fig. 4.1 Lamarck's phylogenetic diagram from his *Histoire Naturelle des Animaux sans Vertèbres...* (1815).

three *grades* of organisation – *animaux apathiques*, lacking a nervous system (according to Lamarck), *animaux sensibles*, with nerves but still largely automata, and *animaux intelligents*. Thus each line of descent with its root represented a *clade*, while the succession of three grades cutting across them represented the *scala naturae*.

It used to be thought that Lamarck's theory was of little influence, and even little known, in at least the first three decades of the 19th century in both France and Britain. It was discussed at length in volume 2 of the three-volume *Principles of Geology* published between 1830 and 1833 by the great British geologist Charles Lyell (see ch. 7 later). Lyell rejected both aspects of Lamarck's theory, phylogeny and the theory of mechanism, but at about the time that he was writing the *Principles* there was a considerable body of support for Lamarckism in Britain, as has been shown by the recent researches of Adrian Desmond. This support was closely involved with the politics of the medical profession, in which Lamarck was accepted by radical practitioners who were fighting the monopoly of the Royal Colleges of Physicians and Surgeons to award medical qualifications. Thus Lamarck's views, that evolution had occurred, and that the deviations from one or more simple *scalae* were produced by a mechanism ensuring the historical adaptation of animals to their environment, were well known and widely accepted (but not 'respectable') in Britain some thirty years before the publication of the *Origin* by Darwin.

the embranchements of Cuvier

Cuvier's principal contributions to biology were his detailed studies in comparative anatomy, so that more than any of his colleagues he based his classifications on the complete anatomical study of animals and not just their external appearance, and his virtual invention of systematic vertebrate palaeontology. In the latter he used his knowledge of the skeletons of living animals (osteology) to reconstruct the skeletons of incomplete fossil animals. But Cuvier resolutely rejected evolution – any evolutionary change in one part of an animal's anatomy or physiology would spoil the coordination of the whole as a working machine, very important to Cuvier, and adaptive change of the whole was too mystical to be believed.

But he also rejected the *scala naturae* in any form. Cuvier's classification of the animal kingdom was based on his *principle of the correlation of parts*, his concept that every animal is a functional machine adapted to its way of life, together with his *principle of subordination of characters*:

> The separate parts of every being must...possess a mutual adaptation; there are, therefore, certain peculiarities of conformation ...which necessitate the existence of others. When we know any given peculiarities to exist in a particular being, we may calculate what can and what cannot exist in conjunction with them. The most obvious, marked and predominant of these, those which exercise the greatest influence over the totality of such a being, are denominated its *important or leading characters*; others of minor consideration are termed *subordinate*.

Thus Cuvier saw an exclusive hierarchy of characters. *The* leading character, which he finally decided was the conformation of the nervous system, determined the basic body plans of animals. On this basis he divided all animals into four major taxa or *embranchements*, firstly vertebrates, possessing a single spinal nerve cord and a brain, secondly molluscs (squids *et al.*, snails and slugs, the marine pteropods, and bivalves) with scattered nerve ganglia, thirdly segmented animals (worms, lobsters *et al.*, spiders and insects) with two ventral nerve cords and ganglia in each segment, and lastly radiates, a mixed bag of echinoderms (sea urchins, star fish), parasitic worms, corals, jellyfish *et al.*, and simple 'infusoria', all with (according to Cuvier) no organised nervous system. The four embranchements were not related to each other in any way, and apart from vital functions, one should not seek resemblances between them, nor could they be ranged in a *scala*. Each embranchement was then divided into (by coincidence or design?) four classes – for vertebrates, mammals, birds, reptiles, fish – and so on, using more subordinate and thus more variable characters for successively finer divisions – class, order, genus, species.

Thus Cuvier firstly saw his classification as the natural order, secondly used all available anatomical characters, thirdly rejected the *scala naturae* and fourthly saw the characters he attempted to use in classification as an *exclusive* hierarchy. The state of the nervous system was the highest ranking character, because it varied only at the highest taxonomic level, that of the four embranchements, but also because it 'exercised the greatest influence over the totality' of any animal. Within each embranchement, another character would then be used to characterise its four constituent classes and so on for lower ranks: hence the exclusive hierarchy of characters used to produce a classification.

I have noted this rather esoteric idea of an exclusive hierarchy of characters in talking about classification for two reasons, firstly as a way of introducing the important controversy between Cuvier and our third eminent French naturalist, Etienne Geoffroy St.-Hilaire, and, secondly because it will lead us into what is perhaps the most important concept in both taxonomy and evolutionary theory, that of homology.

Geoffroy and the concept of homology

Cuvier's view of the characters of any given species of animal was that every one of them had some adaptive explanation. The 'correlation of parts' was to produce a perfectly adapted animal, created to fit its particular environment. Geoffroy's view was different: the characters of any given animal, while mostly adaptive, were each variants of a common plan which could be traced through all animals. The vertebrae of a vertebrate animal such as a fish were basically the same as the external segmental skeletal units of a lobster. A vertebra and the skeletal unit of a lobster were, to use the modern term, *homologous*. Not only that but Geoffroy claimed that by imaginary transformation one could turn all the features of fish anatomy into lobster anatomy or *vice-versa*. These transformations were not seen as reconstructions of evolutionary history (although Geoffroy accepted evolution late in his career) but demonstrations of how the manifestation of the common plan in any one species of animal was equivalent to that in any other. Many of Geoffroy's transformations, particularly between major groups of animals, were fanciful. It was his applauding the demonstration by two youthful and unknown naturalists that a vertebrate bent back at the umbilicus showed many homologies with a squid that precipitated the acrimonious and long-running debate between him and Cuvier.

But Geoffroy's great positive contribution was the method of recognising homologous structures in different animals, the '*principle of connections*'. Homologous structures might look very different but they could be recognised by their similar relationship to other structures. As an example one of the most remarkable homologues in vertebrate animals is that between the chain of three tiny bones, *malleus* (hammer), *incus* (anvil) and *stapes* (stirrup), that conduct sound from the ear-drum down to the inner ear (where the organ of sound reception is situated) in mammals. The principle of connections plus study of the development of embryo mammals shows that the malleus and incus are the homologues to two bones, the articular and quadrate, of all bony vertebrates with jaws except mammals. The articular and quadrate form the lower and upper component of the jaw joint respectively. Land vertebrate other than mammals (birds, reptiles, amphibians) also have the stapes as an ear ossicle but in fishes it is a bone suspending the jaw joint from the braincase, with the connection between the three homologues retained (Fig. 4.2). This particular case was recognised not by Geoffroy but by an anti-evolutionist called Reichert in 1837.

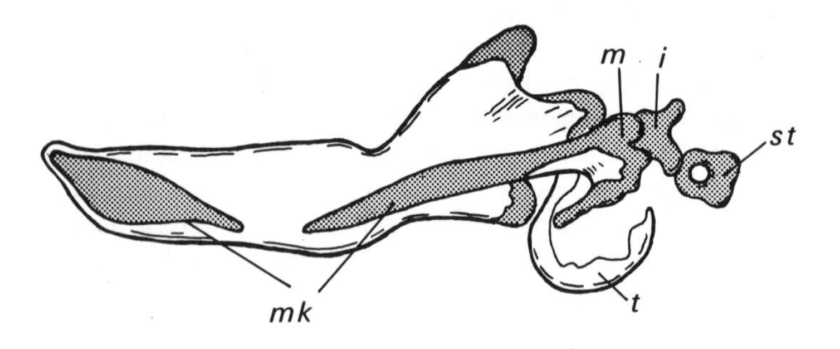

Fig. 4.2 The developing lower jaw skeleton of a mammal (hedgehog): the malleus (*m*) develops as part of the jaw cartilage (*mk*) and articulates with incus (*i*) (the reptilian jaw joint). Stapes: *st*. tympanic (supporting the ear-drum) *t*. (Modified from E.S. Goodrich, *Structure and Development of Vertebrates*, 1930.)

If one sees the Natural Order, at least of animals, as a divergent inclusive hierarchy then Geoffroy's view of the homology of all characters throughout the animal kingdom should lead one to the view that the characters used in classification can each also be ranged in an *inclusive* hierarchy. We can take the example of vertebrate forelimbs. All living vertebrates with jaws (except those, such as snakes, where it has been agreed that they were lost in phylogeny) have a pair of 'pectoral appendages'. The possession of these is then a diagnostic character of jawed vertebrates (gnathostomes). Gnathostomes are divided into two groups, cartilaginous fishes (with 'gristle' instead of bone) such as sharks, skates and rays, and bony 'fishes' ranging from tuna to lung fishes, but also including land vertebrates. They and groups within them are distinguished *inter alia* by a variety of types of pectoral appendages or fins. Land vertebrates (tetrapods) share the feature that the pectoral 'fins' are represented by limbs, i.e. arms with fingers. The basic pattern in tetrapods was for a long time thought to be an arm with five fingers, but recently two of my former research students, Michael Coates and Jenny Clack, have shown that one of the earliest and most primitive fossil tetrapods known, *Acanthostega* (ca. 363 million years), had eight fingers on each hand.*

We can now see that if one takes a character like the forelimb of vertebrates one could draw up a dendrogram representing a hierarchy for just that character. One can then go on to claim that the dendrogram can be interpreted as a 'tree' in just the same way as a dendrogram of animal taxa. We could have done just the same exercise with the homologues of the mammalian ear ossicles which I used as an example above. The dendrogram of homologous 'states' of a character is sometimes known as a 'transformation series'.

homology, analogy and homoplasy

Resulting from the discussion so far we can distinguish two sorts of homology which, by stretching historical accuracy a bit, we can associate with the names of Cuvier and Geoffroy St.-Hilaire. Any vertebrate animal possesses a hollow dorsal nerve cord running through the spine and terminating in a brain at the front. So the nerve cord with brain characterise a taxon, the vertebrates, or more correctly Craniata, one of Cuvier's embranchements. But the nerve cord in any one vertebrate, say a chimpanzee, is homologous with that of any other vertebrate, say a shark. Homology which thus defines a group is *taxic homology*. Possession of three ear ossicles is a diagnostic character of Mammalia; it is also a taxic homology shared by any two mammals.

But using Geoffroy's principle of connections it is possible to identify *transformational homologies*. The malleus of a chimpanzee is the homologue of the articular of a bony fish, the incus is the homologue of the quadrate. The stapes of a chimpanzee, and that of a lizard, is the homologue of the jaw-suspending bone (hyomandibular) of a bony fish. Likewise the wing of a bat, the

* This discovery is the subject of the title essay in the latest of Stephen Jay Gould's books (see p. 3).

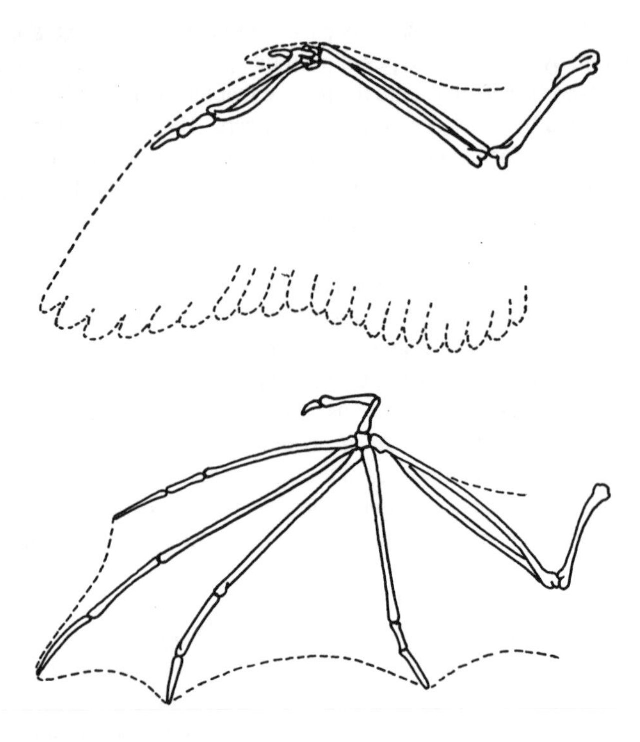

Fig. 4.3 Skeleton of the wings of a bird and of a bat: the aerofoil surface in the bird is formed by feathers, in the bat by skin stretched by elongated fingers.

wing of a bird, the arm of a chimpanzee and the leg of a horse are all homologous as tetrapod forelimbs.

The last example illustrates a new and very important point, which was first clearly codified by the 19th-century British comparative anatomist, Richard Owen, later to be founder and first director of the London Natural History Museum. Let us take the wing of a bat, the wing of a bird and the wing of an insect, e.g. a housefly. They all perform a similar function – are they all homologues of one another? We can eliminate the insect wing. Perhaps some naive and over-enthusiastic follower of Geoffroy might have claimed such a homology, but it can't be sustained on the principle of connections, on similarity of embryological development, or on any other criterion. Because of its similarity of function, i.e. flight, the insect wing is *analogous* to that of a bird, but not homologous. But are the wing of a bat and the wing of a bird homologous? The answer is 'yes and no'! Both (Fig. 4.3) are transformational homologues of the basic tetrapod forelimb, they, or rather the bat and the bird, share the taxic homology of having a tetrapod-type limb, however modified, but they are not homologous *as wings*. There is no taxonomic group that includes bats and birds, all members of which are characterised by wings. So the wing of a bat and the wing of a bird are homologous as tetrapod forelimbs but only analogous as wings. This is to be expected when we have a hierarchy of states of characters, represented by a transformation series.

Owen defined analogue and homologue in the following way in 1843:

> *Analogue*: A part or organ in one animal which has the same function as another part or organ in a different animal.
> *Homologue*: The same organ in different animals under every variety of form and function.

But there is an ambiguity here. Many biologists treat the terms 'analogue' and 'homologue' (also analogy/homology or analogous/homologous) as though they are mutually exclusive, but the wing of a sparrow and the wing of an eagle are both analogous and homologous as wings, similar in both function and derivation. A third term is required for analogous structures which are not homologous: that used was coined by Sir Ray Lankaster in 1870. Such structures are *homoplastic*, the phenomenon is *homoplasy*.

For completeness I must note a third type of homology in addition to *taxic* homology and transformational homology, although it can be regarded as a subset of the latter. The historian of science Toby Appel, in her splendid account of *The Cuvier-Geoffroy Debate*, begins the book by describing the excitement at the debate of the eighty-year-old Johann Wolfgang von Goethe. As well as being one of Germany's greatest poets, he was a good naturalist and philosopher and was probably the first to use the concept, if not the term, *serial* or *iterative* homology. Goethe showed in 1807 that the parts of a flower, each of the sepals (collectively the calyx), petals (corolla), stamens and pistils, could be regarded as a modified leaf, an idea accepted today. They are iterative homologues of the leaf; homologous within a single organism.

Goethe also proposed another case of iterative homology which is not generally accepted today but was very popular for a time: the vertebral theory of the vertebrate skull. The term 'serial homology' is usually used for a linear series of iterative homologues, the segments of a centipede or the vertebrae in any vertebrate animal. But Goethe extended the series to suggest that the skull, also composed of bone (or cartilage in sharks) was elaborated from modified vertebrae. This idea was taken up by a group of mystical anatomists in the early years of the 19th century calling themselves the *Naturphilosophen* (we shall meet them again in ch. 7) and also by Richard Owen.

homology and the natural order

In this chapter I have set out to do two things: firstly to show the contribution of Lamarck, Cuvier and Geoffroy St.-Hilaire to the picture of the Natural Order perceived by Darwin and Wallace; and secondly to explicate the concept of homology in its various forms. I will summarise both. We have seen that the Natural Order was understood by Darwin and Wallace as an irregular inclusive divergent hierarchy, as it is still seen today, and that because of the irregularity, they interpreted the apparent relationships as real, thus proposing that evolution had occurred. Lamarck was among the first to interpret the results of classification as the natural order rather than as the application of correct method, and, of course, interpreted the pattern as phylogeny, the results of evolution. But he also moved away from the simple *scala naturae* to a branching pattern. Cuvier was much more insistent in rejecting the *scala* and his system of embranchements was very influential in establishing the divergent pattern, while Geoffroy, with his principle of connections, put the concept of transformational homology on a rational basis.

But before the results of 'correct' classification could be interpreted as the Natural Order, use of the concept of homology was needed. The notion of taxic homology was the idea that similar structures in different animals (or other organisms) were 'the same' in some special way: a classification that used such structures to unite those animals in a taxon would yield the natural order. Transformational homology goes further and suggests that, using the principle of connections (and also embryology as we shall see) apparently different structures in two or more different organisms can be used not only to put those organisms in the same taxon, but to produce a transformation series. This is in essence a dendrogram based on one character, which could be interpreted as a 'tree' showing the pattern of evolution of that character. The concept of iterative homology is perhaps of less use in taxonomy but is of great importance in suggesting how body plans might have evolved.

With all these concepts in place, we are now in a position to look at current methods of classification and the philosophy that lies behind them.

notes

The standard, (but somewhat inaccurate) English translation of Lamarck's principal theoretical work is **J.-B.-P.-A. de M. de Lamarck**, *Zoological Philosophy* (trans. H.E. Elliot, Macmillan, 1914; recently reprinted by Chicago University Press, 1984). **Adrian Desmond**'s study of the link between political radicalism and Lamarckism in pre-Darwinian Britain is in *The Politics of Evolution...* (Chicago University Press, 1989). **Toby Appel**'s *The Cuvier-Geoffroy Debate* (New York: Oxford University Press, 1987) deals with the views of both as well as the differences between them. The history of comparative anatomy and specifically the concept of homology is dealt with in the classic **E.S. Russell**, *Form and Function* (John Murray, 1916; reprinted in 1982 by Chicago University Press). Recent discussions of homology are **C. Patterson,** in *Problems of Phylogenetic Reconstruction* (eds K.A. Joysey and A.E. Friday, Academic Press [1982] 21-74), **V.L. Roth,** in *Ontogeny and Systematics* (ed. C.J. Humphries, London: British Museum [Natural History] [1988] 1-26) and in my *Classification, Evolution and the Nature of Biology* (Cambridge University Press, 1992).

Clade: the set of all species descended from a single ancestral species.

5: classification up-to-date

Darwinian classification?

> Our classifications will come to be, as far as they can be made so, genealogies; and will then truly give what may be called the plan of creation. The rules for classifying will no doubt become simpler when we have a definite object in view.
>
> (Darwin, *The Origin of Species*... last chapter)

One would have thought that after the publication of the *Origin* an agreed and rational method of classification would evolve rapidly. After all, by 1859 the expected pattern of classification was established, the concept of homology and its implications for taxonomy were generally agreed, and biologists were soon to accept that there should be some sort of relationship between classification and phylogeny. In fact it took over a hundred years before a real attempt was made to develop an objective method for the natural classification of organisms. Then in the 1960s not one but two such methods were devised whose advocates became bitter rivals, and both methods were rejected by traditionalists, at least at first. So what happened?

The history of evolutionary biology from 1859 to the present day is replete with ironies, some of which I will touch on later in this book. The school text-book version is that evolution was accepted as a result of Darwin's and Wallace's work because they for the first time proposed a credible mechanism. What really happened was that Darwin and Wallace persuaded most biologists that evolution had occurred, so that by the end of the 19th century that was the orthodox position, but the theory of Natural Selection did not become orthodoxy until the 1940s and 1950s.

Perhaps the greatest irony is that the general acceptance of evolution after publication of the *Origin* brought about a revival of Lamarckism in all its aspects, particularly among palaeontologists. There are various threads in this revival and it is difficult to disentangle cause and effect, but as far as classification is concerned it produced the malign reappearance of the ancient *scala naturae*. There were various reasons for this but it resulted in palaeontologists seeing their duty as the reconstruction of ancestor–descendent sequences rather than the construction of divergent hierarchies. A large part of their motivation was gathering yet more geological evidence for evolution (see ch. 7). The ideal evidence for evolution would surely be a series of fossils of diminishing geological age and apparent character change, with each specimen representing a descendant of the one before it and an ancestor of the one after. But palaeontologists realised that any particular fossil, or rather the once-living animal it represented, was very unlikely indeed to be the actual ancestor of another fossil or living animal. It was even unlikely, except where the two were very close in time in an exceptionally complete fossil record, that the *species* of one included the ancestor of the other. So statements of ancestry became more vague – 'the reptiles gave rise to the birds' – and reconstructions of phylogeny tended towards John Stuart Mill's second type of *scala naturae*, 'the arrangement of the natural groups into a natural series'. Such strings of taxa were of little use in the production of a hierarchical classification, so that taxonomy continued in a rather *ad hoc* pre-Darwinian manner. This *ad hoc* taxonomy prevailed until the introduction of the two new techniques, *phenetics* and *cladistics* (and still continues today among the unconverted!).

Both phenetics and cladistics produce dendrograms of the type which I described in chapter 1 (pp. 6-8 and Fig. 1.1). That is the dendrograms are dichotomous, divergent, rooted and have the species or other entities being classified only at the lowest rank, i.e. as the external nodes. Internal nodes represent higher ranking taxa which are produced by the method. In both phenetics and cladistics the classification produced is a representation of the dendrogram. But the significance of the dendrogram is very different in the two methods.

phenetics

Originally phenetics was known as 'numerical taxonomy' and developed out of a number of pioneering papers published in the later 1950s and early 1960s. The emphasis was on the use of a very large number of characters, so it is not surprising that the development of phenetics coincided with the development of computer programs capable of manipulating large masses of data. Another principle of phenetics was that all characters should count equally in determining relationships, in the jargon that they should be 'unweighted', or more correctly 'equally weighted'! Cuvier's subordination of characters was rejected. The aim was to achieve a measure of difference between any two species, if it was species that were being classified, a taxonomic distance. If two specimens, representing species to be classified, were similar in all the characters being used the taxonomic distance between them would be zero, or, to look at it the other way round, the 'overall similarity' as it is termed ('aggregate similarity' would be a more accurate phrase), would be one. If the two specimens differed in only one character, i.e. had different 'states' of that character, then the aggregate similarity would be

$$\frac{m}{n} = \frac{n-1}{n}$$

where n is the total number of characters studied and m is the number of matching characters.

Matching characters are those having the same state, as the terms 'character' and 'state' are used by pheneticists. As an example the pectoral appendage (as discussed on pp. 38-41) might be a character while 'fin', 'five-fingered limb' and 'bird's wing' are states of that character; in other words transformational homologues of one another. That, however, would be a very crude example. It is unlikely that a pheneticist would take the whole limb as his character. It would be split up into a whole series of characters, each with two or more states. This brings us to an obvious difficulty. An animal, if animals are being classified, cannot be analysed into a fixed series of independent characters. To take another of our previous examples, do the incus and malleus of mammals (pp. 37-8), which always occur together, represent states of one character (the quadrate–articular complex) or two (the quadrate and articular respectively) or more? But this is a difficulty which also afflicts cladistics, as we shall see.

Most computer programs prefer to deal with binary data – states are recorded as 0 or 1, in the simplest case absent and present respectively. There are then all sorts of rules to deal with multistate characters, e.g. leaves smooth/slightly hairy/very hairy, one/two/three/four/five fingers, to reduce them to binary coding. It is then possible to produce a *character matrix* or *data matrix* which is a table listing all the objects being classified (OTUs – 'Operational Taxonomic Units') along the top and the *n* characters being studied down the side. The column under each OTU then specifies the state of each character, numbers 1 to *n* (Fig. 5.1). With binary characters the distance between any two OTUs can then be calculated easily. The simplest *matching coefficient* can be represented by a two-by-two table (Fig. 5.2). Thus for OTUs **j** and **k**, positive matches (1-1) are represented by *a*, negative matches (0-0) by *d*: *b* and *c* represent mismatches (1-0 and 0-1) respectively. The simple matching coefficient is then

$$S_{sm} = \frac{m}{n} = \frac{a+d}{n}$$

Other coefficients are possible. If 0-0 matches represent the total absence of features and are too numerous, then

$$\frac{a}{n-d}$$

is used, effectively ignoring 0-0 matches.

When coefficients, which are measures of similarity, have been calculated between each OTU and every other OTU, a *similarity matrix* can be drawn (Fig. 5.3). This is like one of the distance tables one finds at the back of motoring atlases. In the latter case the vertical axis and the horizontal axis each have the same list of towns, usually diverging from the corner where the two lists meet. One can then read the distance between any town and any other town at the intersection of the appropriate row and column. This will of course be zero for the same town on row and column. The phenetic matrix has OTUs instead of towns and either similarity coefficients or distance measures (which I haven't discussed) in the cells. In fact most computer programs are simply fed with the data matrix so that one is not required to draw a similarity matrix, but it will be inherent in the program.

n	OTUs				
	A	**B**	**C**	**D**	**E**
1	0	1	0	0	1
2	0	1	1	1	0
3	1	0	0	1	0
4	0	1	0	1	1
5	0	0	1	0	0

Fig. 5.1 A phenetic (or cladistic) character matrix, showing the states of 5 characters (1–5) in each of 5 OTUs ('terminal taxa' in cladistics) A – E.

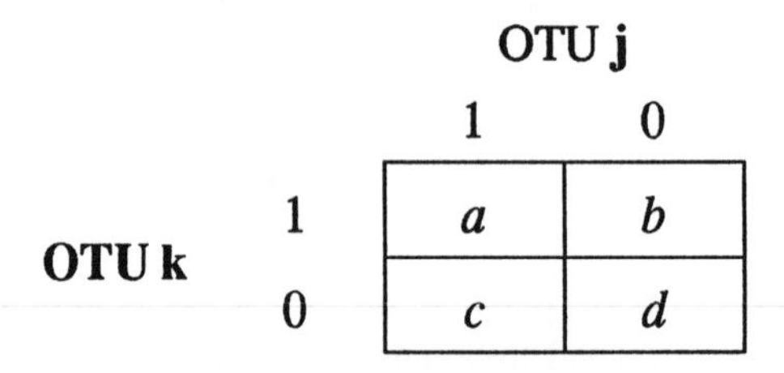

Fig. 5.2 Table of frequencies of matches and mismatches of positive (1) and negative (0) character states between the OTUs **j** and **k**.

OTUs	OTUs				
	A	**B**	**C**	**D**	**E**
A	X				
B	2.4	X			
C	1.5	3.7	X		
D	2.6	5.6	3.5	X	
E	5.1	4.9	1.7	4.1	X

Fig. 5.3 Similarity matrix for five OTUs A–E.

The next stage, whether in the computer program or not, is the clustering procedure, whereby the measures of similarity or distance are turned into a dendrogram. Probably the best way to understand how this works is to take a very simple example indeed. Supposing one has only two OTUs, and one is using only two characters, whose states in the OTUs are being studied. It would then be possible to represent the positions of the OTUs with respect to these characters on a simple two-dimensional graph (Fig. 5.4). The axes of the graph each represent states of a character; the position of each OTU with respect to the two characters is then plotted on the graph. If coding is binary then the only possible position for an OTU or each axis is zero or one, but if distance measures are used, then the state is a continuous variable. On a simple graph, drawn to scale, one can then measure the 'Euclidian distance' between the OTUs with a ruler, or, for those who

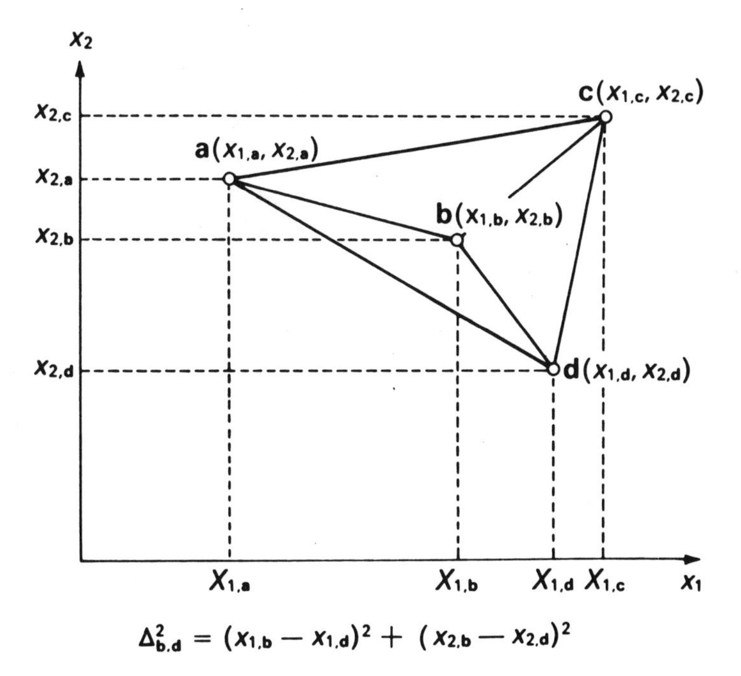

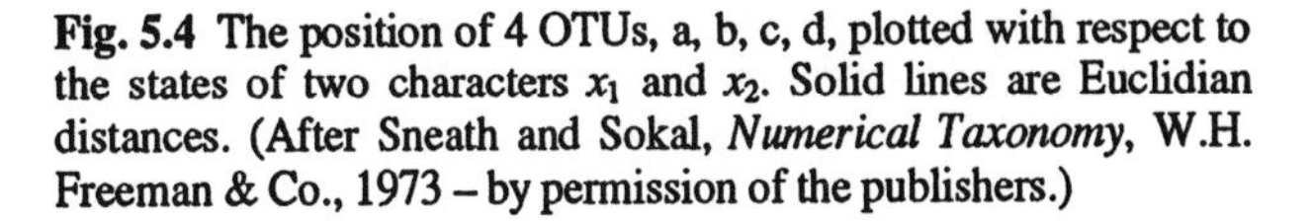

Fig. 5.4 The position of 4 OTUs, a, b, c, d, plotted with respect to the states of two characters x_1 and x_2. Solid lines are Euclidian distances. (After Sneath and Sokal, *Numerical Taxonomy*, W.H. Freeman & Co., 1973 – by permission of the publishers.)

remember their school geometry, calculate it using Pythagoras. Now suppose one is using three characters, a three-dimensional graph is used, which is still possible to envisage and even to draw (Fig. 5.5). But what about *n* characters (where *n* is more, usually many more, than 3)? Theoretically each OTU is plotted on a *n*-dimensional graph or in other words they are distributed in hyperspace. They will not be uniformly distributed, even if their distribution is random. Random distribution even in two dimensions, say of earthworms in a field, always results in clusters; uniform distribution would require some causal explanation, probably that worms repel one another. Similarly random distribution of OTUs in hyperspace will produce clusters and would thus yield a classification even if there were no natural order, 'no taxonomic structure in the data,' but pheneticists can test for this.

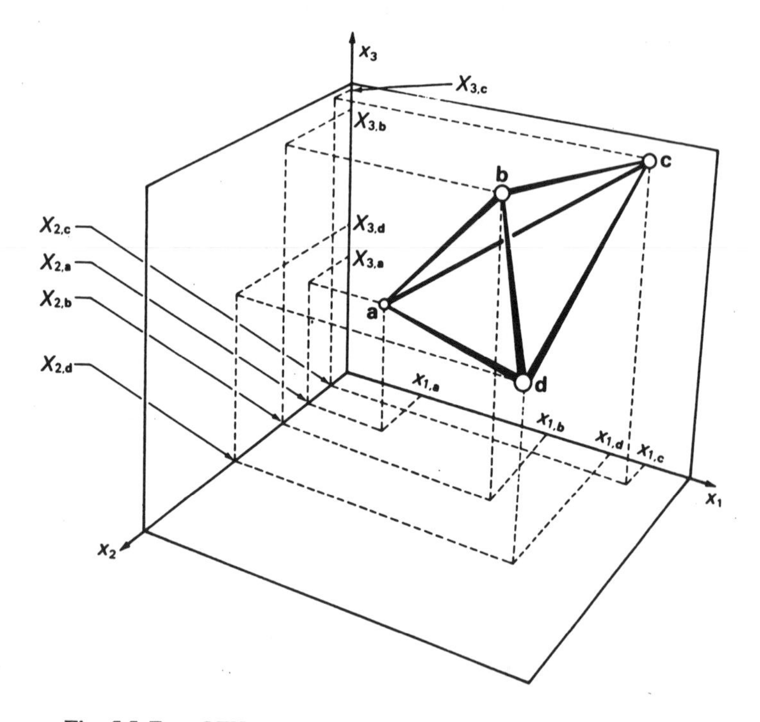

Fig. 5.5 Four OTUs on the axes of 3 characters x_1, x_2, x_3. (After Sneath and Sokal, *Numerical Taxonomy*, W.H. Freeman & Co., 1973 – by permission of the publishers.)

Given the OTUs in hyperspace, the computer searches for the pair which are closest together (have the smallest Euclidean distance). These are then eventually linked on the print-out 'phenogram' by an internal node whose vertical distance from them is a measure of that Euclidian distance. The computer then searches for the next closest pair, treating the previous pair as a single point. If several pairs thus become united as a cluster of a number of OTUs then the cluster is treated as a single point and so on until every OTU has been included. On the resulting phenogram the phenetic distance between any two OTUs is then represented by the vertical distance from either to the nearest internal node uniting them (Fig. 5.6). It is a measure of goodness of fit on the final phenogram, to compare distances of pairs of OTUs on the phenogram with those on the original similarity matrix.

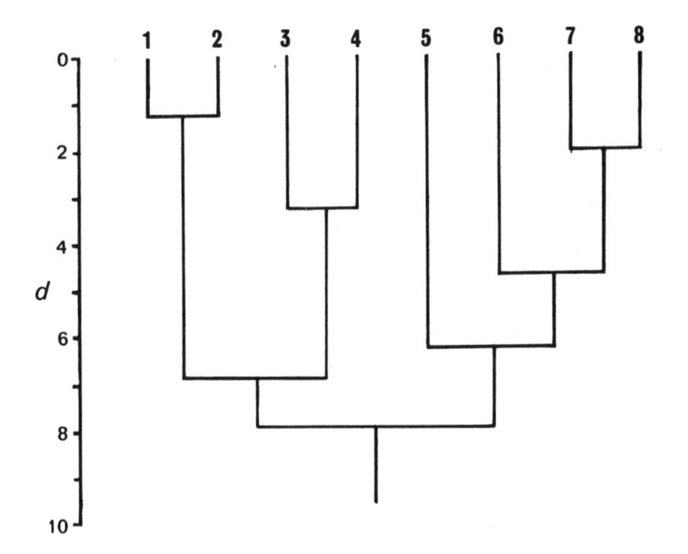

Fig. 5.6 A phenogram for the OTUs 1-8 (*d* is a scale of phenetic distance).

The important question from the point of view of evolutionary theory is to ask what the phenograms produced by pheneticists represent. The pheneticist's main claim is to objectivity. Given the data in the character matrix all the subsequent procedures follow an agreed method with no intuitive component, so that any taxonomist should get the same result. This claim is somewhat vitiated by the fact that there are many formulae for arriving at coefficients of similarity or distance and many clustering programs. As Robert Sokal, one of the founders of phenetics, has said, 'Numerous association coefficients appropriate for binary data are described in various reviews.... Many of these have been applied only rarely (often just once in the original paper proposing their use)'.

There are, however, generally-favoured methods. A more important objection is that a phenogram cannot be interpreted as a tree (see next section). It is important to evolutionary theory to understand why.

phenograms and trees

In the last chapter, I introduced the concepts of homology, analogy and homoplasy (pp. 39-42). Many cases of homoplasy are obvious from observation. Thus, as we saw, the wing of a bat and the wing of a bird are analogous *and* homoplastic *as wings*. This can be corroborated by looking at the structure of their skeletons (see Fig. 4.3). The flying membrane (*patagium*) of a bat's wing is supported by all four fingers leaving a large thumb free. In birds there is still a (very reduced) thumb but two fingers are missing and the other two are partly fused and do not extend to the wingtip. In this case the *patagium*, consisting of flight feathers, has its own stiffening represented by the shafts of the feathers. As I have said, any good pheneticist would analyse these wings as a series of characters with contrasting states.

But some examples of homoplasy are less easy to detect. A good example is in the comparison of the skull of a wolf with that of the (now almost certainly extinct) marsupial 'wolf' of Australia and Tasmania, whose closest relatives were other pouched mammals, the group including kangaroos, koalas and wombats. In the days when comparative osteology was taught routinely in zoology degree courses, it was a standard trick to ask students to identify one skull or the other. The two are remarkably similar in overall shape and in many details of the teeth. Most of the resemblances are analogous and homoplastic, but would be recorded as similar character states in phenetics. The effect of this would be to override the divergent pattern of the natural hierarchy. The skull features of the marsupial wolf would not be distinguished, or weighted differently, from the characters which unite it first with the Tasmanian devil and 'native cats', and then with kangaroos and koalas. In some respects then phenetic classification is like that of Linnaeus, it yields a divergent hierarchy by an approved method, but it is not taken *a priori* that the natural order is a divergent hierarchy.

There is another even more important reason why a phenogram should not be interpreted as a tree. So far I have distinguished between homologous and homoplastic characters, but it is also important to distinguish two sorts of (taxic) homologous characters. Suppose one is studying two very closely related species of animal, so closely related that it is concluded that *of living species* they are more closely related to one another than either is to any other species. To borrow a term from cladistics (see below) they are '*sister species*'. A clear example would be the two living species of elephant. A whole series of characters, long trunk, large flapping ears, tusks, huge molar teeth, are *uniquely* shared by the Indian and African elephant. But they share many more characters which are not unique to them – both give birth to live fully-formed young (they are placentals), both have milk-producing females, a four-chambered heart with a left aorta and a diaphragm (they are mammals), both have limbs rather than fins (they are tetrapods), a skull and vertebrae (they are vertebrates) and so on.

But of all these characters only the uniquely shared ones are useful in grouping the Indian and African elephant as sister species. All the evidence showing the mammalian nature of both is of no help in deciding between an Indian elephant/African elephant pairing and an Indian elephant/rabbit pairing. The mammalian characters become useful, however, when it is decided that the Theria, mammals giving birth to live young, are the sister-group of the egg-laying mammals, thus uniting both groups as the class Mammalia. Thus characters or character states, if genuine homologues, can be thought of in three ways: characters unique to a species or group, for example the molar tooth pattern or much larger ears of the African elephant, characters uniquely shared by two sister-species or sister-groups, and characters irrelevant at the ranks being considered because they distinguish or unite groups at a higher rank. *Thus if the natural order is a divergent hierarchy, then there is a corresponding hierarchy of characters.* In some cases there will be a hierarchy of states of one character – transformational homologues of one another. In other cases separate characters will be used.

cladistics

In saying 'will be used' I have slipped from talking about phenetics to talking about cladistics. While phenetics owed its origin to the independent work of several groups of people, notably Robert Sokal and colleagues, then at the University of Kansas, Arthur Cain, then at Oxford, and Peter Sneath at the University of Leicester, cladistics is almost always credited to one man, the German entomologist Willi Hennig. But Hennig's rationale for the invention of cladistics was not the coincidence of the hierarchy of taxa and the hierarchy of characters; what he was attempting to produce was *Phylogenetic Systematics*, the title of the 1966 English version of his book. Hennig reasoned, firstly (and traditionally!) that classification should be based on evolution and, more particularly, the pattern of phylogeny. He further reasoned (but not traditionally) that the only events in evolutionary history which were all or nothing were the events of speciation or cladogenesis (see pp. 91-2) – hence 'cladistics'. The duty of the taxonomist was therefore to reconstruct the pattern of cladogenesis, which could then be represented by a dichotomous dendrogram, a 'cladogram'. He was not dogmatic about every speciation event being the splitting of one species into two, although some of his early disciples (and I use the word advisedly!) were, but hunting for sister-groups was the most rigorous method.

Thus every species, if it were recognisable at all, would have one, or, much preferably, more than one, unique character. Any given species would then have a sister species and the two would be uniquely united by one or more characters. According to Hennig those latter characters would be diagnostic of the common ancestor of the two species, so that every hypotheses of sister-species status would automatically imply a hypotheses as to the nature of the uniquely shared common ancestor of the pair. In a Hennigian cladogram every internal node represents a hypothetical common ancestor; it also represents uniquely shared characters of the two groups descendent from it. Later, in the 1980s many cladists were to reject the hypotheses of ancestry as unnecessary and concentrate on the hierarchy of characters. So the essence of cladistics was to search for sister-groups, defined by unique characters, then take the group so formed and search for *its* sister-group, which could be a species or any taxon of higher rank, again looking for uniquely-shared characters, until all the species or groups had been incorporated into the cladogram.

But all was not so simple – assuming that you think that was simple! What of homology and homoplasy, and additionally the danger of using a higher-ranking character by mistake? To see the proposed solutions to these problems let us take the simplest problem of classification that a cladist could tackle. Suppose that we have three species A, B *and* C, each represented by specimens which are, of course, the subjects of our investigation. The exercise is to determine that two of the species, say A and B, are more closely related to one another than either is to C. For our conclusion to be reliable we have first to eliminate from consideration characters or states which apply to a higher rank than that of the taxon uniting A, B *and* C. This is done by a procedure known as 'outgroup comparison'.

If, of A, B, and C, a character is shared by two of them to the exclusion of the third, one wants to know if that character is also found in any species (or larger group) outside ABC. If it is then it must characterise a group of higher rank than (A+B+C) and therefore cannot be used to identify a sister-pair within ABC. This outgroup criterion is most reliable if the outgroup (say D) is the sister-group of (A+B+C). Obviously then one has to know something of the taxonomic position of ABC before the criterion can be applied. Knowledgeable readers may detect an 'infinite regress' looming, but usually a satisfactory outgroup or even sequence of nested outgroups can be identified. After that one might think that all would be plain sailing. A and B would have one, or preferably more than one, uniquely shared characters and the problem would be solved: the correct cladogram is:

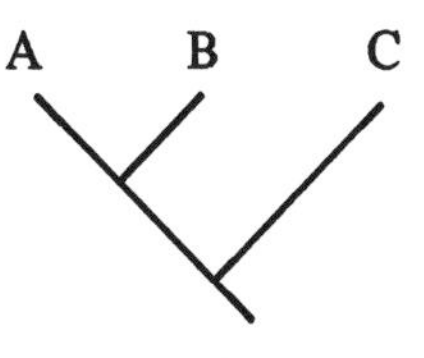

Unhappily and frequently not so! – homoplasy rears its ugly head. There is no trouble with obvious cases of homoplasy. The bird's wing and the bat's wing are so obviously homoplastic as wings in overall structure that no one would allow any similarity to override the obvious fact that the two creatures belong to different classes judged by every other significant character. But what if, in our ABC three-taxon test, there are, in addition to the AB characters, some apparent AC and/or BC characters? Then a very controversial principle is invoked, that of '*parsimony*', an odd use of the principle of economy or parsimony of hypothesis, usually attributed to the 14th-century philosopher, William of Ockham ('Ockham's razor').

In his original book, Hennig had not evolved a clear method of dealing with incongruent characters other than to suggest that contradictory characters be studied more carefully. The principle of parsimony has been developed since then to be (in theory) one of the cornerstones of cladistic practice. My proviso 'in theory' probably stems from my bias as a vertebrate palaeontologist. In attempting the low rank classification of vertebrate fossils, there are frequently so few characters, often of doubtful validity, that the use of parsimony is unrealistic. For the 'principle of parsimony' consists of nothing more than a simple vote: if apparent uniquely-shared AB characters outnumber apparent AC or BC characters then the AC and BC characters are deemed to be homoplastic. It must be obvious that the reliability of this judgement depends on the validity, independence from one another and number of the characters used. The first two are subject to the same worry that I referred to in talking about phenetics – how can one atomise an animal specimen, representing a species, into a series of independent characters of equal importance? In fact, of course, the validity, independence and number of characters used depends on the luck of the draw and the skill of the taxonomist. With living organisms it is probable that the latter will prevail, with fossils there will be a greater component of the former. If challenged on parsimony cladists will reply, with some justification, 'starting from first principles, what other criterion could one use?'

notes

Most of the discussion of taxonomic methods is in specialist journals, notably *Systematic Zoology* and *Cladistics*. The standard text on phenetics is still **P.H.A. Sneath** and **R.R. Sokal**, *Numerical Taxonomy...* (Freeman, 1973). The most accessible text on (phylogenetic) cladistics is **E.O. Wiley,** *Phylogenetics...* (John Wiley,

1981). **Colin Patterson** gives a succinct summary of ('transformed') cladistics in the journal *Biologist* 27 (1980) 234-40. (See also Patterson 1982 and my 1992 – notes to ch. 4.)

Cladogram: a branching diagram (*dendrogram*) representing the hierarchical grouping of species, or higher taxa, based on uniquely shared characters.

Phenogram: a dendrogram based on grouping by aggregate similarity.

Speciation: the splitting of one species into two or more, resulting in some degree of reproductive isolation between the 'daughter species'. (The process is less clear cut in plants than in animals.)

6: a sceptical interlude

the horror of hybrids

> [Hybrid species] can appear as unresolved basal taxa, as sister taxa to one of the putative parents, or fully or partially collapse the cladogram.... Furthermore, we have no 100 per cent certainty of distinguishing between multiple speciation, living ancestors, reticulation, or homoplasy, and thus we seem to have reached the limits of cladism....
>
> (C.J. Humphries in *Advances in Cladistics* vol. 2 [1983] 89-103)

Chris Humphries is a plant taxonomist and is one of the principal advocates of cladistics among botanists. The epigraph to this chapter is taken from the proceedings of the second meeting of the Willi Hennig Society, which was founded in 1980 as a breakaway group from the (American) Society of Systematic Zoology. The Hennig Society was founded to allow cladists to talk amongst themselves unperturbed by the dissenting views of pheneticists and traditionalists, yet by the second meeting cladistic botanists were already worrying about a phenomenon that they saw as a threat to cladistic procedure – speciation by hybridisation. But a case can be made that the existence of hybrid species threatens not just cladistics, but the *a priori* assumption underlying the theory of evolution. In order to tackle this alarming suggestion I will pose two questions based on our discussion so far:

1. What is the theory [that evolution has occurred] for? Or, more formally, of what *explanandum* is the theory of evolution the *explanans*; i.e. what corpus of 'facts' and/or lower-level theories does the theory purport to explain?

2. What is its success in this respect? – (a) is it an adequate explanation of the *explanandum*, and (b) are there, or could there be, any rival theory or theories which do an equally satisfactory or more satisfactory job?

I hope that the reader will agree that I have answered question 1 in the preceding chapters of this book. The theory of evolution is to explain the Natural Order. The natural order, as perceived by the theory's most successful proponents, Darwin and Wallace, is an irregular divergent hierarchy. Shortly before their joint publication in 1858 and the publication of the *Origin* in 1859, this perception was reinforced by the concepts of homology, analogy and what we now call homoplasy, with rules for the recognition of homology. It was further reinforced by the assertion of divergence by Cuvier and to some extent the later Lamarck, and of irregularity principally by Darwin and Wallace themselves.

Turning to question 2, I hope that the reader will further agree that the answers are (a) – 'yes', and (b) – 'no'. 'Community of descent', in Darwin's words seems so obviously the correct answer to the apparent relationships of classification, that any rejection of that explanation must surely be due to ignorance, stupidity or prejudice. Nor can I think of any scientific theory which would explain the apparently contingent nature of the divergent, irregular hierarchy better.

There is another important point here. The advocates of 'special creation' as a rival theory to evolution, mostly American christian fundamentalists, cannot claim the same *explanandum* of the pattern of classification. I do not propose a detailed discussion of 'special creation' in this book: I am striving to explain science and expound logic. But for those interested I will add some references at the end of this chapter.

In talking about cladistics in the last chapter, I talked principally about cladistic *methods*. I did, however, say that Hennig's original aim was to produce a rigorous method for reconstructing the pattern of phylogeny, or at least its cladogenetic component. Classification was then to be based on the resulting cladogram. But this is at odds with the fact that logically and historically the pattern of classification is prior to the theory of evolution. Evolution should not therefore be taken *a priori* in devising methods of classification. Yet traditional taxonomists and Hennigian cladists ('phylogeneticists') do just that. Pheneticists don't (although they do in fact assume evolution), but they do not claim to be reconstructing the natural order. Only one group of contemporary taxonomists claim to be reconstructing that order, but do not invoke evolution *a priori*. I mentioned on p. 54 that in the 1980s many cladists rejected the idea that the internal nodes of the cladogram represented hypotheses of ancestry and saw the cladogram solely as a representation of a natural hierarchy of characters. This led eventually to the rejection of *a priori* evolution (but not a rejection of belief in evolution). This group of cladists, notably Gareth Nelson and Norman Platnick at the American Museum of Natural History and Colin Patterson at the British Museum (Natural History), became known to their opponents, but not to themselves, as 'transformed' or 'pattern cladists'. They saw their stance as a rational development of Hennigian cladistics: their opponents, both within and without the cladistic fold, saw it as a betrayal. The methodology of Hennigian cladists and transformed cladists is virtually the same (but see ch. 7), but their stated aims are different. Hennigians think they are reconstructing the pattern of phylogeny: transformed cladists think they are reconstructing the natural order. Incongruent characters eliminated by parsimony (see our 'three-taxon test' pp. 55-6) are interpreted as 'convergent evolution' by the Hennigians, but simply as initial mistakes of interpretation by the transformed cladists.

The fact that the latter group reject incongruent characters as 'mistakes' demonstrates that to them the Natural Order is a real phenomenon – it is 'out there' waiting to be discovered. Furthermore they have stated their belief that the natural order is a divergent hierarchy. What they did not do, initially at any rate, was clearly to justify their non-phylogenetic stance by citing the *explanandum/explanans* argument, the logical priority of classification to evolution. This justification was in fact offered to them *on passant* in a paper by me on the use of parsimony published in 1982, but was more fully developed by a philosopher, Ronald Brady, in their own journal *Cladistics* ('The International Journal of the Willi Hennig Society') in 1985. There is, however, an oddity in Brady's account. He sets out the *explanandum/explanans* argument with admirable clarity and says that in many cases studied a stable and robust hierarchical pattern is produced by cladistic methods, suggesting in those cases that the natural order is a divergent hierarchy. But then Brady goes on to say –

> About the groups that we cannot resolve we may say nothing, for our knowledge is insufficient to make a judgment as to whether these groups are somehow different or simply have not been resolved *as yet*. Our claims to knowledge must rest upon the well resolved groups, and these are most interesting to the investigator.

Even apart from its echoes of Wittgenstein – 'Whereof one cannot speak, thereon one must remain silent' – this seems to me very peculiar. Having established the logical priority of the natural order to phylogeny, is Brady then saying that we only know the nature of the natural order as a divergent hierarchy in those groups of organism for which we have a successful cladistic classification? Or, to put it another way, is he advising that one should not infer by induction (or at least propose as a hypothesis) that the natural order of *all* organisms is a hierarchy from those cases where there is good reason to think so? Here then is the paradox of transformed cladistics. In rejecting the *a priori* assumption of phylogeny they have things in the right logical order, but have no extrinsic evidence that the natural order is a divergent hierarchy, because the only access they have to that hierarchy is by a method of classification which assumes its existence *a priori*. By using outgroup comparison in classification, they are assuming that the natural order is hierarchical, and by using parsimony, they are assuming that that hierarchy is divergent. Note also that this implies a hierarchy of characters and character states which is logically prior to and of the same form as (isomorphic to) the natural order.

Now is the time to reintroduce the botanical cladists' worry about hybrid species. It is known that new plant species can arise by hybridisation, resulting from crosses between distinct parent species. The parent species need not be sister species of one another. Thus a 'tree', in the cladist sense (p. 63), which included the hybrid would not be wholly divergent: it would be '*reticulate*'. But cladistics cannot yield a reticulate cladogram. If the hybrid favoured one parent species rather than the other it would probably appear on the cladogram as sister species to that parent. If it favoured neither, and the parent species were true sister species, then the taxonomist would get an 'unresolved trichotomy', three species which would not yield to a three-taxon test. If the parent species were not sisters the results of cladistic analysis would be unpredictable.

Is hybridisation a serious taxonomic problem? For plants, almost certainly yes. Botanists vary widely in their estimates, but it has been estimated that nearly 60% of flowering plants have at least one hybrid species in their ancestry. For animals, probably, no, although there are hybrid animal species. One famous example is *Rana esculenta*, the edible frog of gourmet fame and probably the commonest frog species in Europe. Its status was elucidated only in the early 1970s. It is remarkable in that in most populations it is not a fully independent species but must co-exist with one or other of its parent species, the pool frog or the marsh frog. In reproduction the edible frog behaves like the complementary species to whichever parent species is present, producing pool frog gametes (eggs or sperm) when mating with the marsh frog, and *vice-versa*, thus reconstituting the hybrid.

the pattern of phylogeny

So does the undoubted existence of hybrid species refute the status of the natural order as a divergent hierarchy? Answering that question is beset with all sorts of complications. This is not a case where a single statement – 'there are hybrid species' – has the potential to falsify a 'universal statement' (i.e. a scientific theory) according to the original canons proposed by the philosopher of science, Sir Karl Popper. The possibility of such irrevocable falsification is not now generally accepted (even by Popper), but even if it were, the true phylogeny is not a universal statement of the form 'of all points in space and time it is true that...' but a single historical and thus contingent entity. But, furthermore, 'there are hybrid species' simply tells us something about the *pattern of phylogeny*, having accepted that evolution occurs. What does it tell us about the Natural Order, which is logically prior to phylogeny? I suppose that if the *explanandum/explanans* principle is rigidly maintained, one could answer 'nothing', although that answer flies in the face of common sense. But it is possible, I believe, to undermine the absolute priority of *explanandum* over *explanans*, or pattern of classification over that of phylogeny, or cladogram over tree (to put it three ways). My colleague Tim Smithson and I are among those who claim to have done so for a particular case.

In order to explain our reasoning I must introduce another term in addition to 'cladogram' and 'tree', as used by cladists. The three were expressed thus by Niles Eldredge of the American Museum of Natural History in 1979:

> [A] *cladogram*...is a branching diagram depicting the pattern of shared similarities *thought to be evolutionary novelties*...among a series of taxa.
> [A] *phylogenetic tree* [is] a diagram...depicting the actual pattern of ancestry and descent among a series of taxa.
> [A] *phylogenetic scenario* is a phylogenetic tree with an overlay of adaptational narrative.
>
> (*emphases mine*)

Note that for a transformed cladist the phrase 'thought to be evolutionary novelties' would have to be removed from the definition of a cladogram. Thus a logical order of priorities can be set up, which if we include the original taxonomic characters used to construct the cladogram, goes **characters → cladogram → tree → scenarios.**

Eldredge says of scenarios:

> Scenarios are inductive narratives...concocted to explain how some particular configuration of events...took place. The hallmark of such narratives is the analysis of the adaptive significance of evolutionary changes in size, form and structure...they are mostly fairy tales constructed of a maze of untestable propositions concerning selection, function, niche utilisation, and community integration, and alas, do not generally represent good science.

Eldredge is a palaeontologist, so he was thinking of scenarios mostly in terms of reconstructing the biology of organisms found as fossils. This reconstruction will include 'soft anatomy' (the parts that don't fossilise), physiology and behaviour including the interpretation of the function of structures in the fossil(s), hence functional anatomy, ecology and biogeography. But scenarios are also popular amongst 'neontologists', which is what we palaeontologists call every other sort of biologist. They are particularly popular amongst those students of animal behaviour who wish to interpret every feature, anatomical, behavioural or whatever, of their study organisms as adaptive (see ch. 12).

The particular scenario which Smithson and I invoked was in the course of a review of the relationship of tetrapod vertebrates to particular groups of bony fish. The review was an attempt to refute a previous one by four colleagues (Rosen, Forey, Gardiner and Patterson – all 'transformed cladists') in which they concluded that the lungfishes (Dipnoi) were the sister-group of tetrapods, not only if extant ('living') groups are considered, but even if, in addition, all fossil groups are taken into account as well. The orthodox position is that members of a fossil fish group (the osteolepiforms) are closer. The standard cladistic technique is to produce a cladogram of the extant groups first and then introduce any fossil groups in relation to it. Following strict cladistic procedure we concluded, in agreement with Rosen *et al.*, that the lungfishes were the *living* sister-group of tetrapods, but disagreed with them by saying that, considering *all* relevant fish groups, the osteolepiforms were the tetrapod sister-group.

But then we went further than that and questioned whether our strict use of cladistics (and that of Rosen *et al.*) had produced the correct extant-group pairing. The only valid uniquely-shared characters uniting lungfishes and tetrapods that we could find were soft-anatomy characters all concerned with the fact that living lungfishes are lung-breathing and have a suitably modified blood system. They included a heart partially divided to separate oxygenated and de-oxygenated blood, separate circulation to the lungs, and a flap closing of the entrance to the lungs to exclude food and/or water.

The three living genera of lungfishes, from Australia, Africa and South America respectively, all live in freshwater and in the latter two cases can breath by lung alone without using their reduced gills. The gill cover in all three is reduced in size relative to that in other bony fish. Lungfishes back to Coal Measure times (about 300+ million years ago) were also freshwater or swamp forms with reduced gill covers, but those from the earlier Devonian period (ca. 400-360 million years ago) were mostly marine (90% for Early and Middle Devonian: 400-377 million years). Furthermore, they had large gill covers and, where known, a fully-formed gill skeleton. We inferred, therefore, that the features apparently uniting extant lungfishes and tetrapods had evolved separately in the two groups, they were analogous and homoplastic: the marine lungfishes of the Devonian would not have needed to breath air as when a swampy lake dried up or became anoxic. While they probably had lungs, or at least their homologue an air bladder, it is very unlikely that they had the air-breathing specialisations of Coal Measure and more recent lungfishes. If we are right then our scenario about the respiratory anatomy and physiology of fossil lungfish refutes the lungfish-tetrapod shared characters and the inferred cladogram.

Note that we suggested an evolutionary scenario. If correct we have established a 'feedback loop' from evolutionary scenario to cladogram: attempts to establish the natural order must take reconstructed evolution into account. The *explanandum/explanans* principle is thus breached. But if that is the case we can take the evidence of hybrid species to show that the whole of the natural order is not a divergent hierarchy. The most extreme conclusion that one could reach from this is that we know nothing of the natural order. If that were the case, evolution would have lost its *explanandum* and the theory that evolution has occurred would be unnecessary!

Fortunately I do not think that such an argument can be sustained, although it would be reassuring if there were some extrinsic evidence of the hierarchy. The way in which we can climb back from the brink is to note that a 'transformed' cladogram is merely a pattern of branching. If one then goes on to interpret it as the pattern of phylogeny (i.e. as a tree), it represents only the pattern of cladogenesis. Anagenesis, evolutionary change in anatomy etc., of which we have evidence (ch. 7), is not represented except incidentally as producing differences between diverging species. Now imagine a phylogenetic tree in which the branch lengths between nodes can differ from one another and are each a measure of the 'quantity' of anagenesis or phyletic evolution that has occurred between the nodes at each end. In that respect this 'anagenetic tree' would be like a phenogram produced by phenetic clustering, but there would be two important differences. Firstly the tree would not be manipulated to make all the species-to-be-classified (terminal taxa) appear at the same level, and secondly homologous and homoplastic characters, and characters significant at different ranks would not be conflated as in phenetics.

computer cladistics

To produce such a tree one would need some measure of 'evolutionary distance' between species. This could be based on anatomical etc. characters, as in normal cladistic classification, or on data from biochemistry and molecular biology. To go into detail of the techniques in either case would be beyond the scope of this book, but they are sufficiently important to our theme to persuade me to attempt an outline.

One technique for producing a tree with significant branch lengths originated with a plant taxonomist, W.H. Wagner. Shared unique characters were used in the tree, so that by retaining the branching pattern but ignoring the significance of the branch lengths, or redrawing, the pattern could be regarded as a cladogram. The method is best explained by a diagram (Fig. 6.1) in Wagner's original style. ABC are the three species to be classified. The numbered semi-circles represent unique characters. The node at zero represents the root of the tree, a 'hypothetical ancestor' having only significant characters shared by A, B *and* C. Then A and B have six uniquely-shared characters marked by a node at 6. In addition A has two characters unique to itself giving a total distance from the root of 8 units, while B has 4 giving a total distance from the root of 10. C has no characters uniquely shared with A or B, but has 7 unique characters. This Wagner scheme can then be turned into a cladogram identical to that in our three-taxon test (pp. 55-6).

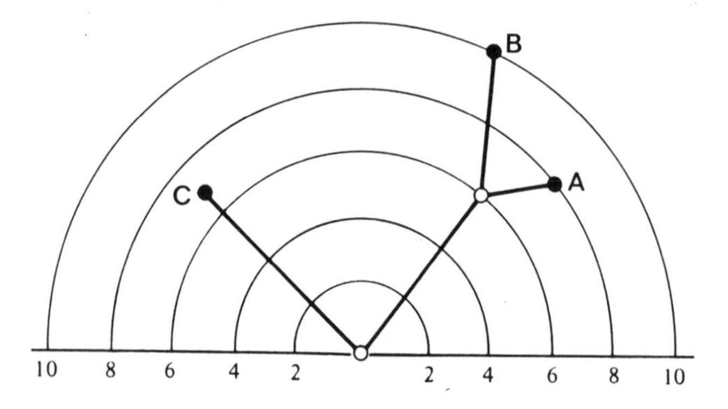

Fig. 6.1 Wagner ground-plan-divergence method for three taxa ABC (see text). (Modified after Wiley, 1981.)

But it is also important that the principle of parsimony can be introduced into the Wagner scheme. The pattern with the highest number of uniquely shared, as distinct from unique, characters will have the shortest total aggregate branch length. Outgroup comparison is also involved to determine the suite of characters represented by the root (zero) node. Wagner analysis was therefore ideal as the basis for computer cladistics and was so used by J.S. Farris of the State University of New York and his colleagues. They have produced computer programs which could handle large numbers of taxa and large numbers of characters, as in phenetics, but produce rooted trees by specifying the characters of an outgroup in the data matrix. So Wagner programs produced trees whose branch lengths were a measure of the 'amount' of evolution (i.e. anagenesis or phyletic evolution) that had occurred.

If every character in a Wagner tree was of equal 'value', that is in some sense represented a unit of evolution, then total branch lengths from root to species or other taxa being classified would be a measure of rates of evolution and *ceteris paribus*, i.e. with no other way of eliminating all homoplasy, parsimony in the form of minimum branch length would be the only rational choice. In molecular biology one has such unit characters. They are the units (amino acids) that make up protein molecular that carries the hereditary program, the famous 'double helix' of DNA (deoxy-ribonucleic acid).

molecular taxonomy

There are two ways of investigating the linear structure of either protein or DNA. Distance measures or sequencing. Distance measures, from immunology in the case of proteins, and from the pairing of DNA from two different organisms in the case of DNA, yield the equivalent of the phenetic classification of characters (which works on calculated distances). In the case of sequencing of proteins, however, the actual characters are available. All proteins are made up of single or branched chains of amino acids (often with other non-amino acid units), but only twenty different amino acids ever appear. In the case of DNA there are only four possible bases (each paired with another particular one of the four) like an alphabet with only four letters. If the bases are regarded as the letters, each word has three letters. This allows the possibility of 4^3 (= 64) words, more than adequate to code for only twenty possible amino acids. Thus a length of DNA present in one of an organism's chromosomes codes for a corresponding length of protein with each 'triplet' of bases ultimately being responsible for one amino acid. Because there are sixty-four different possible triplets but only twenty amino acids to be coded form, there is considerable 'degeneracy' in the code: usually but not always the third base in a triplet can be any one of the possible four without affecting the result.

We can see that given the sequence of amino acids in a protein, easier to obtain than a DNA sequence, the corresponding DNA sequence can be reconstructed at least for the first two bases of each triplet. One can then compare homologous proteins, and by inference their DNA, between any two organisms. Favourite proteins for animals have been the oxygen-carrying proteins, haemoglobin, which gives the red colour to vertebrate blood, and myoglobin which colours red meat. The simplest sort of evolutionary change that can happen to DNA is a 'point mutation', substitution of a base by one of the remaining three. So from the protein comparison one can reconstruct the minimum number of mutations separating the two organisms.

Given those data for a number of organisms a classification can be produced. It could be produced using phenetics, as the pioneers did, but to do so would be to underuse the data. The second method is to use parsimony, that is a cladistic technique. This is particularly associated with Walter Fitch of the University of California and Morris Goodman of Wayne State University, Detroit and his colleagues. Another approach is to use 'likelihood' in which it is assumed that mutation of any base is a random event in time, like the breakdown of a radioactive atom, and thus to produce a dendrogram which is closest to the prediction of randomness for the data. This approach was pioneered by Joe Felsenstein of Washington University and also developed by Martin Bishop and Adrian Friday in Cambridge. Both methods have their advocates, but both give (with different assumptions) a tree with branch lengths proportional to evolutionary distance. So they are methods of classification in which the 'characters' used are less subjective than the case when using anatomical characters. Furthermore, as well as yielding a pattern of cladogenesis, the *explanandum* of evolutionary theory, they also yield measures of evolutionary rate and thus of anagenesis. They thus embody not only the *explanandum*, but also something about phylogeny which is *a posteriori* to the theory of evolution (*s.s.*).

notes

Two 'texts' of (American) creationism are **D.T. Gish,** *Evolution? The Fossils Say No!* (San Diego: Creation – Life Publishers, 1979), and **H.M. Morris**, *Scientific Creationism* (general edition, 1974), from the same publishers. There are a number of books that set out to refute creationism. Among the best is **Philip Kitcher**'s *Abusing Science, the Case against Creationism* (MIT Press, 1982; Open University Press, 1983).

There is a textbook account of hybrid speciation in **D. Briggs** and **S.M. Walters**, *Plant Variation and Evolution* (Cambridge University Press, 1984). The *explanans/explanandum* distinction was pointed out in **A.L. Panchen**, *Zoological Journal of the Linnean Society* 74 (1982) 305-28, but much more fully developed in **R.H. Brady**, *Cladistics* 1 (1985) 113-26. **Eldredge** sets out the distinction between cladograms, trees, and scenarios in *Phylogenetic Analysis and Paleontology* (eds J. Cracraft and N. Eldredge, Columbia University Press [1979] 165-98). Our lungfish scenario is in **A.L. Panchen** and **T.R. Smithson**, *Biological Reviews* 62 (1987) 341-438. Wagner analysis is explained in Wiley's text (notes to ch. 5) and molecular and anatomical techniques discussed in **C. Patterson** (ed.), *Molecules and Morphology in Evolution* (Cambridge University Press, 1987).

Chromosomes: elongated bodies in living cells (within the nucleus if present) that carry the hereditary program as a linear series of genes. During cell division take up microscope stains strongly.

DNA (deoxyribonucleic acid): the active principle in chromosomes, consisting of elongate helical molecules embodying the genetic code and thus representing a series of genes.

'*Degeneracy*': the situation where more than one sequence of bases in DNA molecule codes for the same (amino acid) product.

Mutation (genetic): spontaneous origin of an hereditary character, or the resulting character. Now usually an error in replication of the hereditary material or its results. Can apply to a single DNA base ('point mutation') or part of whole chromosome.

Phyletic evolution: evolutionary change in taxonomic character(s) or ancestor–descendent sequences of organisms over time (usually equated with anagenesis, ch. 1).

7: the evidence for evolution

evidence from anatomy and embryology?

> The reason why it is believed that the enormous variety of life has evolved from simple ancestors is based on three main sources of evidence. They are the evidence from *fossils*, from *affinities* between organisms and from the *geographical distribution* of organisms.

The quotation is from an English school textbook *Advanced Biology* (Third Edition) by Simpkins and Williams, published in 1989: 'advanced' because it is aimed at school students taking the General Certificate of Education, Advanced Level, the national examination usually required for university entrance. Later under 'affinities' the authors go on to say:

> All living organisms have common features. Some have features which are virtually identical to those of another organism. In other words, they have a close *affinity* to each other and for this reason they are classified in the same group. Among the features which indicate that living organisms are related by descent are *anatomical, physiological, embryological, biochemical*, and *behavioural* affinities.

The reader will immediately recognise (I hope!) that the 'features' referred to by these authors are the data, the characters, used in classification. Anatomical features are the stuff of orthodox classification (including phenetics and cladistics) and we looked briefly at the use of immunology and other biochemical and molecular techniques (pp. 67-8). It is also easy to see that physiological and behavioural characteristics could also be used in classification: Simkins and Williams note the common features of chimpanzee and human behaviour; but embryology needs further comment.

In chapter 3 I mentioned the German *Naturphilosophen* in talking about Goethe's use of the concept of what we now call iterative homology. Goethe's combination of artistic and scientific endeavour is similar to the aim of *Naturphilosophie* to penetrate the mystery of nature. Its leading practitioner, Lorenz Oken (1779-1851), was preoccupied by a theory of cosmic development, and by man as the microcosm of which the universe was the macrocosm. All animals were degenerations of the human form – a *scala naturae* read from top to bottom, with universally traceable homologies like those of Geoffroy St.-Hilaire. Because of this the stages of development of the human embryo were represented by the sequence of animals on the *scala*. To some extent this it true if one considers only vertebrate animals. The human embryo does go through a stage when it has gill pouches, but not the perforated gill slits, like those of a fish; the internal partitions of the human heart do develop in such a way that there is some resemblance to that of an amphibian and then a reptile, and the human embryo does have a tail, which is subsequently lost, or rather reduced and 'internalised'.

This 'double parallel' between human development (*ontogeny*) and the *scala* was subsequently elaborated into a triple parallel, notably in a notorious anonymous book, *Vestiges of the Natural History of Creation*, first published in 1844. The author was Robert Chambers (1802-71), co-founder of the publishing house, but he managed to retain his anonymity until after his death. This first book in English on evolution combined the ideas of Oken and other *Naturphilosophen* with those of Lamarck and introduced the fossil record as the third parallel, with (at least vertebrate) fossils appearing in the geological sequence in the same order as the stages of embryology and of the *scala*. This parallel was the basis of the theory of 'recapitulation' later made famous by the German evolutionary publicist Ernst Haeckel (1834-1919) with the aphorism 'ontogeny recapitulates phylogeny'.

Which is not strictly true – it was another naturalist brought up in the tradition of *Naturphilosophie*, Karl Ernst von Baer, who demonstrated in 1828 that (vertebrate) embryos do not simply climb to different definite levels on the *scala* in their development but diverge from one another.

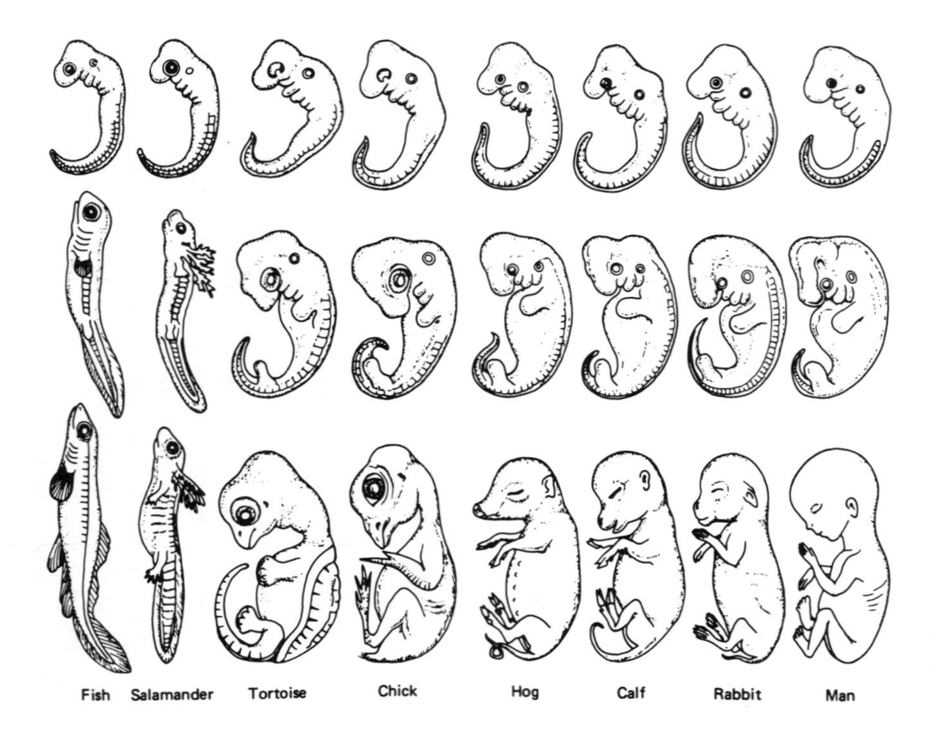

Fig. 7.1 Comparison of vertebrate embryos at three stages of development to illustrate von Baer's laws (from W.D. Stansfield, *The Science of Evolution*, Collier Macmillan, 1977 – by permission of the author).

> Each embryo of a given species instead of passing through the stages of other animals departs more and more from them. Fundamentally therefore, the embryo of a higher animal is never like [the adult of] a lower animal, but only like its embryo.
>
> (*von Baer's third and fourth laws*)

But here (Fig. 7.1) we have another irregular divergent hierarchy, and one, like a taxonomic tree, on a time base. The time, however, is in the realm of individual development not phylogeny. According to von Baer, the early embryos of two different species share most of their characters but diverge by acquiring more and more specialised characters as they develop. High ranking characters develop first, successively lower ones later. This is not universally true; a hen's egg has many special features (adapting it to be free of standing water) that develop before features characterising all tetrapods, including those for example of spawn-laying frogs. But the generalisation is sufficiently true for transformed cladists to attempt to use the hierarchy of embryological characters rather than 'outgroup comparison' to polarise characters (see p. 55). So the pattern of development in comparative embyology stands revealed as yet another facet of the *explanandum*, data for classification rather than evidence for evolution.

vestigial organs

One can then ask, 'are there any data from comparative anatomy and embryology which are acceptable evidence for evolution' – and this can be answered 'yes, vestigial organs'. The human tail is such a vestige, so are the remnants of the hip girdle and hind limbs of whales, which are entirely embedded within the body, and the similarly reduced hind limbs of primitive snakes such as the boa constrictor, which appear externally only as a pair of spurs either side of the vent. Likewise many cave-dwelling and deep-sea animals have very reduced eyes and yet in other ways are closely similar to light-living relatives. In the great baleen or 'whalebone' whales, such as the blue whale, teeth are absent. The animals feed by filtering sea water to retain the plankton through a series of long narrow plates of baleen made of keratin, the stuff of hair and finger nails. Yet their embryos erupt apparently functionless teeth which are soon lost. Even more striking is the work by Kollar and Fisher described by Stephen J. Gould in his famous essay *'Hen's teeth and horse's toes*'. In most vertebrates teeth develop from potential jaw tissue, which yields the internal dentine of the tooth, interacting with the ectodermal tissue overlying it, which yields dental enamel. Kollar and Fisher cultured mouse jaw tissue from the molar region with chicken ectoderm and produced simple teeth unlike the molars of mice. These were 'hen's teeth'. The genetic potential to produce dental enamel and to induce dentine is still there in chickens, although the last toothed bird in the fossil record occurs at about the same time as the last of the dinosaurs, some 65 million years ago. In all these cases it is difficult not to believe that the vestigial structures in the extant animals were represented by fully-formed ones in their ancestors. Anagenesis or phyletic evolution has occurred.

fossil evidence

The evidence from the fossil record is also mostly of anagenesis. The usual assumption is that the fossil record is *the* evidence for evolution, that the history of the evolution of taxa is manifest in the rocks. A sobering statement at the other extreme is that of David Kitts, a historian and philosopher of science, in 1974:

> ...fossils, by themselves tell us nothing; not even that they are fossils.... When a paleontologist decides whether or not something found in the rocks is the *remains* of an organism he decides in effect, whether or not it is necessary to invoke a biological event in the explanation of that thing. Sometimes a judgement is made without hesitation because the object in question resembles some part of a living organism.... But.... There is no structural feature which serves to distinguish fossils from entities of every other kind....

In other words fossils can only be identified as remnants or traces of once-living organisms by reference to and comparison with extant organisms. To interpret the fossil record as evidence for evolution a number of additional observations and inferences are required. I will summarise the sequence of conclusions arising from these and then discuss them:

1. that fossils (as now characterised) are evidence of once-living organisms and not some parallel phenomenon;

2. that fossils, in their stratigraphic setting (i.e. their reconstructed

sequence in the rock strata) constitute a historical record;
3. that extinction is a real phenomenon;
4. that there is progression in the fossil record, i.e. that the record shows a succession of faunas and floras (assemblages of animals and plants) and other organisms through geological time, such that individual species in the record are taxonomically related to but different from those which precede and succeed them;
5. that despite known major events, the fossil record is not marked by any catastrophic event that appears to have eliminated the whole biota (fauna, flora *et al.*); and further that the first appearance of fossil taxa, and their disappearance, do not always coincide with known major extinction events;
6. and thus that apparent progression in the fossil record is more plausibly interpreted as phyletic evolution (anagenesis) than as 'catastrophism'.

Although nowadays it is generally accepted, even by 'creationists', that fossils, and most obviously such fossils as vertebrate skeletons and mollusc shells, are the remains of once-living organisms, that has not always been the case. During the 17th and 18th centuries there was a vigorous debate between advocates of the modern view and those that believed that fossils represented a parallel creation, never alive but simulating living things. The matter is often said to have been settled by Steno (Niels Stensen), a Dane, writing in the 1660s. He began by recognising that the common 'tongue stones' (*glossopetrae*) of northern Mediterranean strata were similar in external appearance *and* internal (histological) structure, with dentine and enamel, to the teeth of living sharks – '*the principle of sufficient similarity*': Gould. Steno then reasoned that the fossil glossopetrae found within a rock sediment had impressed their form on the sediment without themselves taking up the form of pre-existing cavities within the rock – '*the principle of moulding*'. The fossils must therefore once have had an independent existence, predating the consolidation of the sediments in which they were found. That should have settled the matter, but there were in 17th- and 18th-century terms two rational objections to extending Steno's views to all fossils. Firstly many fossils were known even then which had no known living counterparts, implying extinction, and, secondly, patently marine fossils were known from mountain strata. Extinction seemed an impious theory, implying imperfection in God's creation, and there was no agreed solution to the problem of location, which is now fully explained in terms of earth (tectonic) movements.

My next three points, fossils as a historical record (2), the reality of extinction (3), and progression in the record (4) were largely accepted soon after the beginning of the 19th century and can be discussed together. Historical geology depends on two principles, taught to generations of students as the *'principle of superposition'* and the *'principle of correlation'*. The first is delightfully simple but all important. In an undisturbed sequence ('section' to geologists) of sedimentary rocks (those formed from pre-existing particles as in sandstone or mudstone, or by precipitation from solution or by chemical action as with limestone), the oldest will be at the bottom. The sequence in the section will thus represent the order of deposition from bottom to top. The second principle, that of correlation, concerns the determination of the relative ages of sedimentary rocks in different localities. If a sedimentary stratum in a section in one part of the world has the same or a closely similar fossil biota to one from somewhere else in the world, then they are of approximately the same age. The converse does not apply, two sediments from different localities with very different biotas are not necessarily of significantly different ages; they could, for instance represent contemporary biotas, one marine and the other terrestrial or freshwater.

For the principle of superposition to be not just true but useful it was necessary to assume that the accumulation of sediments over the earth, and their erosion and other processes, have taken place over an immensely long period of time, rather than say deposition in rapid sequence during and after Noah's Flood. This former conclusion became orthodoxy by the end of the 18th century and has been corroborated in this century by radioactive dating of sediments by direct or indirect means. So the fossils within those sediments are fragments of the history of life.

Extinction had to be accepted to counter the hypothesis, that members of fossil groups not now known to be extant had undiscovered living members in some corner of the world as yet not adequately explored. This was believable for relative small marine creatures such as the ammonites, the spiral shells of squid-like creatures. But at the end of the 18th and beginning of the 19th century large fossil mammals, mammoths, ground sloths and others, were discovered in the Old and New World. It was unlikely that such large terrestrial creatures would be found alive on increasingly well explored continents. Even less is this the case with the dinosaurs, everybody's favourite fossils, found in the 19th and 20th centuries. So extinction is a reality: when some commonly found fossil species disappears from the record it is likely that it had ceased to exist. If in a rock section each major stratum has its own distinctive suite of, for example, ammonite species (representing an ammonite 'zone') as is the case on the north Somerset coast, then the distinct species of one zone are assumed to have become extinct before the base of the succeeding zone. But unless the biota of the world has been rapidly and regularly diminishing since the beginning of the fossil record all the new fossils characterising a zone must have come into existence at or shortly before the time represented by the base of the zone. There is change or 'progression' as it was termed in the fossil record. This does not necessarily imply direction or 'improvement', just that fossil species have come into existence, and almost always become extinct relatively soon after, throughout the three-and-a-half thousand million year history of life on earth.

But there are two rival interpretations of progression in the record, 'catastrophism' and evolution. They were fiercely debated from the beginning of the 19th century onwards.

Acceptance of the reality of extinction was very much due to the work and the advocacy of Cuvier and his colleague Alexandre Brongniart. Cuvier had been responsible for the description of some of the dramatic large mammal fossils recognised as extinct and had suggested that only some great catastrophe could have killed off these huge well-adapted creatures. He and Brongniart then embarked on their famous study of the sediments of the Paris basin. Before that study Cuvier had envisaged only a single 'world anterior to ours' and separated from it by one great catastrophe; then he and Brongniart discovered what appeared to be a series of catastrophic events in strata extending from a time soon after the extinction of the dinosaurs (which they did not know about) to the present. It was also they who shared the honour with William Smith (1769-1839; an English civil engineer concerned with investigation of rock strata for cutting canals), for enunciating the principle of correlation. The Paris strata showed an alternation of marine and freshwater beds with appropriate mollusc and other fossils. But two successive freshwater beds, separated by what was interpreted as a marine incursion, had different faunas. The rational explanation seemed to be that the fauna of the lower bed was extinguished by the catastrophic event, then the following freshwater bed was repopulated from elsewhere. Cuvier did not envisage universal catastrophes as he was strictly a cautious empirical scientist, but some English geologists, notably William Buckland (1784-1856), Reader at Oxford, did – and at first interpreted the most recent as Noah's Flood.

How could Cuvier and Brongniart's eminently reasonable 'catastrophic' interpretation be countered and 'progression' be interpreted as evolution? If some fossil species lived through presumed catastrophes ('extinction events' in modern terminology) and their first and last appearances did not coincide with such events, then it would not be possible to draw a line across the whole biota of the world at a given time and declare that at that time the old biota had been replaced by a new one. Any *systematic* appearance of progression in the fossil record, a succession of fossils apparently demonstrating directional change, must then be attributed to evolution.

The absence of total extinction at any given time was demonstrated for the same geological period as that of Cuvier and Brongniart's work by Charles Lyell, recorded in the *Principles of Geology* (1830-3), although his aim was to name and characterise a series of geological epochs in the period concerned. These epochs (correct technical term!) he labelled, from the oldest to the youngest, Eocene, Miocene, Older Pliocene, Newer Pliocene, and they were characterised not by having a diagnostic fauna in each case, but by the percentage of extant species in their fossil mollusc fauna: the Eocene with about 3%, the Miocene ca. 20%, the Older Pliocene 33-50+% and the Newer Pliocene ca. 90%. Thus catastrophes or no, some species still living had survived the whole period and there was a net increase in these throughout. There is, however, an irony in using Lyell's work in support of evolution. He rejected evolution until late in life and at the time of writing the first edition of the *Principles* rejected progression. The biota at any time in earth history was determined purely in relation to climatic and other environmental factors. If past environmental conditions recurred –

> Then might those genera of animals return, of which the memorials are preserved in the ancient rocks of our continents. The huge iguanodon might reappear in the woods, and

> the ichthyosaur in the sea, whilst the pterodactyle might flit again through umbrageous groves of tree ferns.

Given then that 'progression' in the fossil record is more logically interpreted as due to evolution, how should one present particular fossil sequences as evidence for evolution? I suggest that there are two types of evidence – phyletic evolution of taxonomic characters and phyletic evolution of taxa themselves.

fossils and phylogeny

As an example of the first we can take the ear ossicles of mammals discussed in chapter 4 (pp. 37-8). Of the three ossicles, the stapes (stirrup) inserts into a 'window' in the inner ear and conducts sound in all tetrapods (but its homologue in most fish suspends the jaw joint). The other two, incus (anvil) and malleus (hammer) complete the chain and also conduct sound in mammals, the latter being inserted in the eardrum, but are homologues of the quadrate and articular respectively which form the jaw joint in all vertebrates with jaws, other than mammals. There is developmental evidence: in the embryos of primitive mammals, such as the hedgehog, the incus and malleus develop in continuity with the developing jaws, and the newly-born kangeroo, tiny and undeveloped, still has a reptilian-type jaw joint when it begins to suckle.

Now we can turn to the very full fossil record of mammal forebears. Members of the mammal clade, traditionally known as 'mammal-like reptiles' but more correctly 'stem-group mammals' or even 'reptile-like mammals', first appear in the Carboniferous Period, some 315 million years ago, while the first 'true' mammals, defined by their distinctive new jaw joint, appear near the end of the Triassic Period, about 210 million years ago. In the intervening period more than 100 million years, the stem-group mammals flourished mightily and hundreds of species of fossils have been found, the earliest particularly from North America, then from the (former) USSR and finally with a world-wide distribution, but particularly from South Africa and South America. Through this vast collection of fossils one can trace a transformation series towards the mammalian condition; enlargement of what is the only jawbone of extant mammals, the dentary, reduction of the quadrate and articular, which become loosely articulated in the skull, and change in many correlated structures.

Note that in this case of evolution of a character complex I have invoked the triple parallel of the *Naturphilosophen* and the '*Vestiges of Creation*'; an implicit linear transformation series – embryology – fossil record. Evidence of the evolution of characters can legitimately be cast in this linear form; primary evidence of the evolution of taxa should not be.

As an example of the latter let us take a cliché example, the evolution of the horse. All extant horse species, domestic and wild

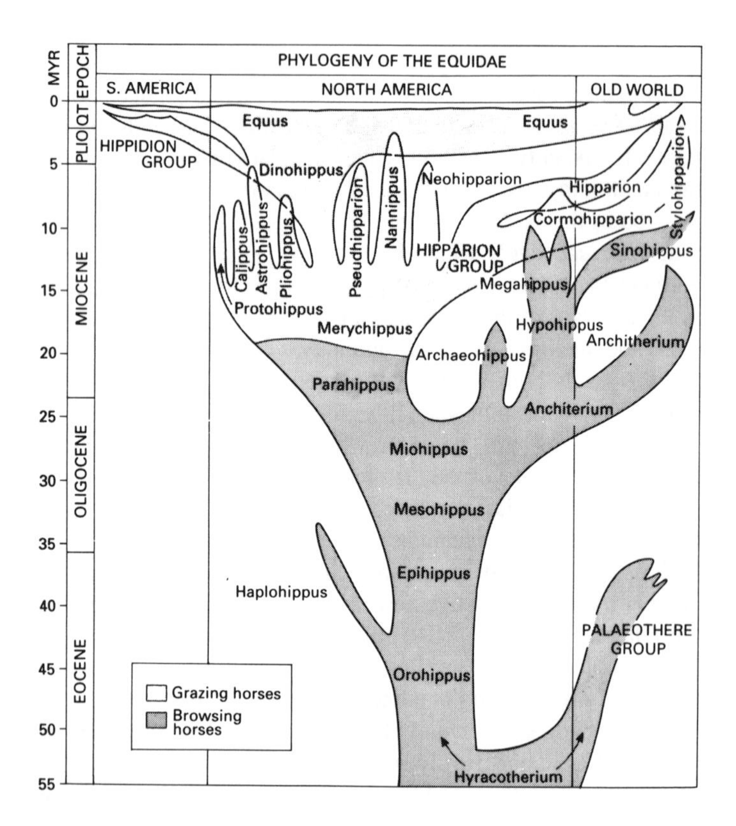

Fig. 7.2 Phylogeny of the horse family (Equidae) in time and space. (After B.J. MacFadden, *Paleobiology* 11 (1985) 245-57 – by permission of the Paleontological Society.)

horses, asses and zebras, are members of the same genus *Equus*. The earliest known member of the horse family, usually known as '*Eohippus*' but correctly *Hyracotherium*, a little creature 'about the size of a fox terrier' (see Gould's amusing essay on that phrase), flourished in the early Eocene epoch, about 55 million years ago. Early accounts of horse evolution, and later textbooks, had a simple *scala naturae* account of the origin of the advanced characters of extant horses. Among those characters are a single toe and toenail (hoof) on each foot, long persistently-growing premolar and molar teeth, with self-sharpening vertical plates of dentine, enamel and cementum in their crowns, and a bar of bone in the skull behind the 'orbit' in which fits the eye.

But the fossil record of horses shows a much more complex pattern (Fig. 7.2). There was not just a single evolutionary line from *Eohippus* to *Equus* but a complex bush with, notably, a great radiation of primitive browsing horses, with low-crowned molars, coexisting with early grazing horses in which the complex high-crowned teeth are taken to have evolved. How then should one present such a fossil record as evidence for evolution?

In our example of the evolution of the ear ossicles the fossil record yielded stages in the transformation from the primitive reptilian condition to the mammalian one, but if I were to give a detailed account of all the fossils demonstrating the transformation series I would *not* be making the claim that they could be arranged in an ancestor–descendent sequence, although it is implicit that, if we had specimens of every species of stem-group mammal that ever existed, such a series could be traced. Similarly if we had every species of the horse family, Equidae, there would be (presupposing evolution for a minute) an ancestor–descendent sequence from '*Eohippus*', or something closely related to it, to *Equus*. But as I said in chapter 5 (p. 45) it is improbable that known fossils from any major taxon contain a complete sequence from most primitive to extant, and we know (don't we?!) that the Natural Order is not such a *scala naturae* but (all together!) an *irregular divergent hierarchy*. So what we do is invoke a different type of double (or if possible) triple parallel: cladogram, dated phylogeny (and if possible) embryology.

Figure 7.3 is a cladogram showing the classification of just seven horse genera, based on the states of just four characters, body size, number of toes on the fore-foot (*Eohippus* has four), high-crowned teeth ('hypsodonty'), and absence/presence of a post-orbital bar. The greatest age of any specimen of each genus is shown. Now if the cladogram is regarded as a tree, each internal node from the bottom up represents the hypothetical ancestor of all the branches descendent from it, with their uniquely shared character(s) indicated. Thus R, N, P and Q all have a post-orbital bar and at least partial hypsodonty. But note also that each internal node can be assigned a minimum age, that of the oldest genus descendent from it: the common ancestor of R, N, P and Q must have been extant *at least* 20 million years ago. So now we have a reconstructed ancestor –descendent sequence from which our discovered genera are presumed to have branched off. We have reconstructed a series of ancestors from *Eohippus* to *Equus* from the more complete series which must have existed, but is unknowable. Corroboration of our

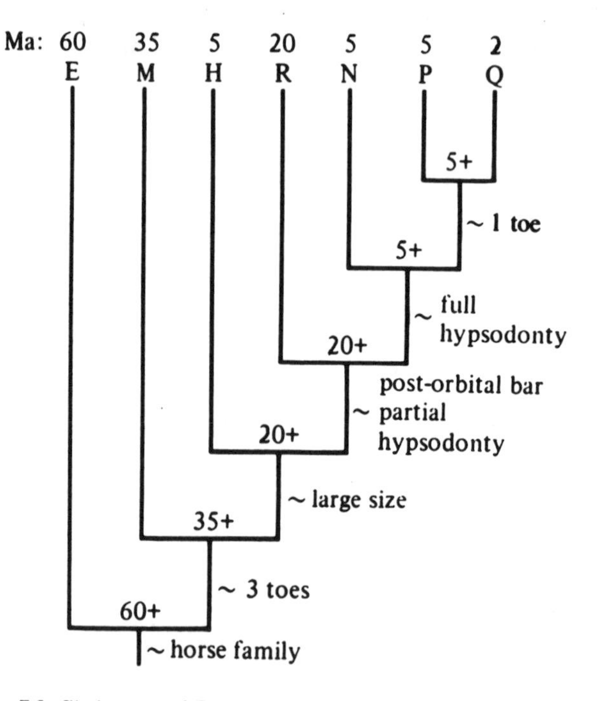

Fig. 7.3 Cladogram of 7 horse genera – E, *Eohippus*; M, *Mesohippus*; H, *Hypohippus*; R, *Merychippus*; N, *Neohipparion*; P, *Pliohippus*; Q, *Equus* – dates of earliest occurrence in millions of years (Ma) along the top; minimum dates of branch points (common ancestors) indicated (hypsodont = high crowned).

reconstruction and, I suggest, powerful evidence of evolution, is the fact (not quite perfect I admit) that the age of the nodes diminishes throughout from ultimate ancestor to extant genus.

geographical evidence

In considering the evidence for evolution to be inferred from biogeography it is possible to list a series of necessary conclusions analogous to those we had for the fossil evidence:

1. the pattern of distribution of animals and plants cannot be explained by environmental factors alone: their history must be invoked;
2. comparison of geographically distant but similar environments makes it improbable that each species was separately created in the environment to which it is most highly adapted;
3. *Ceteris paribus*, the resemblance between any two faunas is inversely related to the width and/or effectiveness and age of the barrier between them;
4. Regions of geographical endemism [see below] can be arranged in a divergent hierarchy of 'regions within regions' corresponding to the taxonomy of their contained species.

Speculation about historical biogeography goes back at least to Linnaeus but it really began as an organised discipline with the French botanist Augustine de Candolle who in 1820 made the distinction between the '*station*' and the '*habitation*' of a plant. The station was an indication of the environment to which the plant was adapted: the habitation the geographical area in which it is found –

> The station of *Salicornia* is in salt marshes; that of the aquatic *Ranunculus*, in stagnant freshwater. The habitation of both these plants is in Europe, that of the tulip tree in North America. The study of station is so to speak, botanical topography, the study of habitations botanical geography.... Stations are determined uniquely by physical causes actually in operation...habitations are probably determined in part by geological causes that no longer exist today.

A number of people before Candolle had noted the contrast between the Old World and the New World fauna and flora. He went further and divided the terrestrial globe into 20 regions of endemism (subsequently 40), i.e. major regions to which part of their flora was

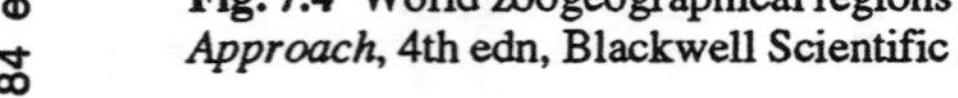

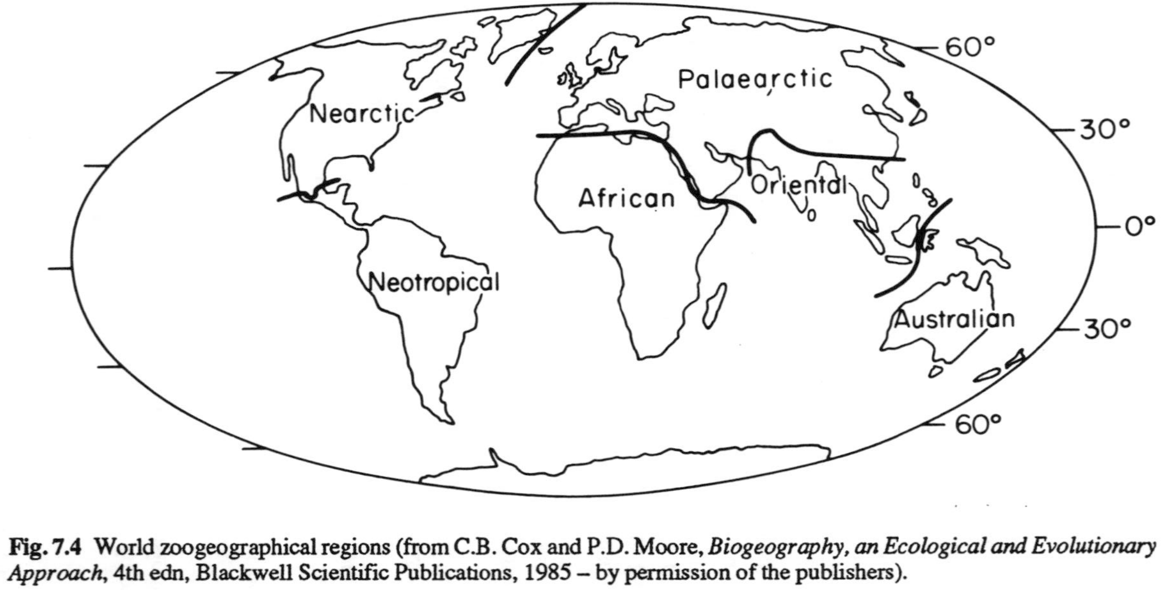

Fig. 7.4 World zoogeographical regions (from C.B. Cox and P.D. Moore, *Biogeography, an Ecological and Evolutionary Approach*, 4th edn, Blackwell Scientific Publications, 1985 – by permission of the publishers).

exclusively confined. With this came the recognition that other plant species were more or less cosmopolitan, and a discussion of their probable pattern of dispersal.

Soon after Candolle's initial work zoologists started to delimit land regions of the globe. Eventually Sclater in 1858 defined six regions based on endemic birds and these were adopted by Alfred Russel Wallace in the *Geographical Distribution of Animals* (1876) and are accepted today (Fig. 7.4). Wallace gives an enormously detailed study of the fauna of each region. In most cases there is an obvious barrier to land animals between them: the Atlantic separates the two New World regions from the Palaearctic and African regions, which are separated by the Sahara and the deserts of the northern Arabian peninsula. Similarly the Oriental region is bordered by the Himalayas. But the nature of the separation between the Oriental and Australian regions was mysterious to Wallace and has only been explained with the acceptance of plate tectonics.

But why should the faunal regions be evidence of past history and thus for evolution? Could there not be subtle differences in the environment of, say, the rain forest of the Amazon region from that of African which explain why the former has jaguars and 'new-world monkeys' while the latter has leopards, 'old-world monkeys', chimpanzees and gorillas? Fortunately (or in fact very unfortunately!) transplantation experiments world-wide have been undertaken fortuitously by our own species. Dingo dogs, rabbits, pigs, buffalo and even camels have been introduced into Australia, flourished mightily and devastated the native mammal fauna of pouched mammals. As Darwin put it in the *Origin*:

> No country can be named in which all the native inhabitants are now so perfectly adapted to each other and to the physical conditions under which they live, that none of them could anyhow be improved; for in all countries the natives have been so far conquered by naturalised productions, that they have allowed foreigners to take possession of the land.

Or again:

> In the southern hemisphere, if we compare large tracts of land in Australia, South Africa, and Western South America, between latitude 25° and 35°, we shall find parts extremely similar in their

> conditions, yet it would not be possible to point out three faunas and floras more utterly dissimilar. Or again we may compare the productions of South America south of latitude 35° with those north of 25°, which consequently inhabit a considerably different climate, and they will be found incomparably more closely related to each other, than they are to the productions of Australia or Africa under nearly the same climate.

Thus it is improbable that the differences between the biotas of similar environments widely separated from one another can be explained by special creation, but more positive conclusions can be reached from such comparisons.

A favourite comparison of Darwin's was between the Cape Verde islands off the west coast of Africa and the Galapagos islands off the west coast of Ecuador, South America. Both archipelagos are tropical volcanic groups, which have thus never been part of the nearest continent, and have a closely similar climate. Notably both lack amphibians and terrestrial mammals, which cannot easily disperse over oceans. But in each case the affinity of the fauna and flora is with that of their respective continental mainland. The Cape Verdes have a bird fauna, notably with several species of kingfisher, derived from that of West Africa, while that of the Galapagos is derived from that of South America as are other elements of the fauna such as the land and marine iguana lizards.

But then one can make another comparison, between the biota of oceanic islands with reference to the nearest mainland (as with the Cape Verdes and the Galapagos), and close continental islands such as the British Isles. The British fauna is principally an impoverished version of that of continental Europe, unique (e.g.) mammals are merely endemic *sub*-species as with field voles. But the famous 'Darwin's finches' of the Galapagos are an endemic sub-family. There is another divergent hierarchy lurking here. Wallace in his 1855 'Sarawak Essay' suggested that the pattern of distribution of animals mirrored the ranks of classification:

> Large groups, such as classes and orders, are generally spread over the whole earth, while smaller ones, such as families and genera, are frequently confined to one portion, often to a very limited district.

He goes on to claim that the degree of difference between the fauna either side of a natural barrier is proportional to the effectiveness of that barrier and to its age.

> When a range of mountains has attained a great elevation and has so remained during a long geological period, the species of the two sides...will often be very different...a similar phaenomenon [sic] occurs when an island has been separated from a continent at a very early period....
>
> In all those cases in which an island has been separated from a continent, or raised by volcanic...action from the sea, or in which a mountain chain has been elevated, in a recent geological epoch, the phaenomena of peculiar groups...will not exist.

Wallace thus saw the possibility of a geographical taxonomy but it took until the mid-1970s, with the acceptance of cladistics, for this to begin systematic development. Vital to that development was the distinction between *dispersal* and *vicariance* as historical explanations of the distribution of animals and plants.

The historical pattern of dispersal from its place of origin of a species, voluntarily or passively, is unique to that species. But vicariance, the origin of geographical barriers, can be expected to affect a whole number of not necessarily closely related species in the same way. An early worked out example of *vicariance biogeography* concerned a genus of trees, *Nothofagus*, the southern beeches. These are known not only as living trees, but as fossil pollen from South America, Australia, New Zealand and (pollen only) Antarctica, but are conspicuously absent from Africa. All these areas are now known to be parts of the great southern continent Gondwanaland, which began to split up with the separation of Africa about 100 million years ago. Given our knowledge of plate tectonics it is possible to draw a dendrogram showing the order of separation of the land masses. It is also possible to draw a cladogram of the large numbers of species of *Nothofagus*. Comparison of these two dendrograms will then show whether the pattern of speciation reconstructed in *Nothofagus* is compatible with the pattern, the order, of the break-up of Gondwanaland. Each could corroborate the other. But it is possible to go further than that. Several unrelated groups of animals have a similar distribution to *Nothofagus* including two families of midges, a group of earthworms, mollusc groups, freshwater fishes and some birds, as well as other plants.

In a perfect world separate cladograms for all these would be congruent with that for *Nothofagus*. The consensus cladogram would then be an *area cladogram* which would have geographical areas instead of species as its terminal taxa. The taxa of each group at all ranks would then be comparable to the characters of normal cladistic classification. In fact this was done for the two midge families. The results are not entirely consistent with the *Nothofagus* cladogram, but they allow a reconstruction of the order of speciation due to vicariance events and its comparison with the geological record.

Vicariance biogeography is in its infancy, but it offers the possibility of the reconstruction of patterns of speciation and their corroboration by reconstructed geological events. Because of the latter, it is not part of the *explanandum* of evolution, and if it achieves any success this can only be because the pattern of the natural order is due to a pattern of speciation, cladogenesis. There is an irony here because many of its advocates, Gareth Nelson, Norman Platnick, the late Donn Rosen, Colin Patterson, Chris Humphries, are 'transformed' cladists anxious to pursue taxonomy without reference to evolution.

notes

I originally attempted logically to set out the evidence for evolution as in this chapter in my *Classification, Evolution and the Nature of Biology*. The relationship between evolution and development, both as science and history of science is discussed in **Stephen Jay Gould**'s *Ontogeny and Phylogeny* (Harvard University Press, 1977). His essay on 'hen's teeth and horse's toes' is reprinted in the book of that title, that on Steno in the same book – 'The titular Bishop of Titiopolis' – and on the... 'size of a fox terrier' in *Bully for Brontosaurus*. The standard historical work on the significance of the fossil record is **Martin Rudwick**'s *The Meaning of Fossils* (2nd edn, Chicago University Press, 1985). For biogeography see **Wallace**'s 1855 paper (notes to ch. 2 in this book), and the historical introduction in **G. Nelson** and **N.I. Platnick**, *Systematics and Biogeography: Cladistics and Vicariance* (Columbia University Press, 1981).

'*Catastrophism*' (W. Whewell): interpretation of the history of life from the fossil record as a succession of different faunas and floras separated by world-wide catastrophic events.

Ontogeny: the development of an individual organism from the fertilised egg to adulthood.

Vicariance: the initiation of speciation (ch. 5) and thus cladogenesis (ch. 1) by the origin ('vicariance event') of a barrier dividing the parent species range.

8: theories of mechanism

> ...I am convinced that Natural Selection has been the main but not exclusive means of modification.
>
> (*The Origin*... end of Introduction)

Under what circumstances would one expect evolution to stop? – I ask this extraordinary question to introduce the reader to the two categories of factors, *intrinsic* and *extrinsic*, that one must take into account in describing and criticising any theory of evolutionary mechanism. Lamarck would have been sure of the answer: it would need a suspension of natural law! For, as we saw in chapter 4, the basis of his theory of mechanism was an innate tendency to evolve, with spontaneous generation of the simplest organisms (*Monas*) and over the generations evolution from *Monas* to humankind, driven by 'subtle and ponderable fluids'. But Lamarck, as a good naturalist, was also preoccupied by *adaptation*, the evident fit of organisms to their environment. This he explained by the 'inheritance of acquired characters' – 'Lamarckism', the 'wrong' theory of evolution as explained to British students (p. 2). In fact starting from first principles 'Lamarckism' would probably seem the obvious solution to the problem of the evolution of adaptations. A similar solution was sketched by Charles Darwin's grandfather, Erasmus Darwin, was allowed a possibility by Darwin himself in the *Origin*, and elaborated by him into the very unconvincing 'provisional theory of pangenesis' in *the variation of Animals and Plants under Domestication* (1868).

Lamarck and 'Lamarckism'

Lamarck propounded the adaptive part of his theory as three laws:

> The production of a new organ in an animal's body, results from a new need which continues to make itself felt, and from a new movement that this need brings about and maintains.
>
> The development and effectiveness of organs are proportional to the use of those organs.
>
> All that is acquired or changed during an individual's lifetime is preserved by heredity and transmitted to that individual's progeny.

A 'first law' preceding these embodies the innate tendency to evolve.

Thus in Lamark's theory the origin of new organs resulted from the response of the animal to environmental pressures: webbed feet were acquired by water birds by the action of stretching their toes while swimming using their feet. There were different levels of response – animals with a well-developed nervous system produced new behaviour which stimulated, over the generations, the production of new organs. In lower animals and plants it was the subtle fluids that responded. The law of 'use and disuse of parts' would have seemed like common sense to an early 19th-century naturalist. The cliché blacksmith's biceps were enlarged by his vocation; his sons (and presumably daughters) would be born with slightly larger biceps than would have been the case if the same man had been a chartered accountant – Lamark's penultimate and last laws in action. The same laws would explain the many blind animals characteristic of totally dark caves as reduction and in some cases loss of eyes as the result of the inheritance of disuse.

In Lamarck's theory then the intrinsic factors which made evolution inevitable were of two sorts: first, the innate tendency to evolve along, or up, the course of the *scala naturae*, and, second, the ability to produce heritable adaptive features in response to the extrinsic factors, changes in the environment, or rather the ecological *niche*, of the organism. Anagenesis, evolutionary change with time, was related to both sorts of intrinsic factor – innate evolution (technically *orthogenesis*) and response to the environment. Cladogenesis, or the splitting of one species into two, presumably was thought to result from different members of species adopting or being forced into different modes of life. But thinking about speciation in a modern way requires thinking in terms of populations of organisms, with the members of those populations differing from one another, as Ernst Mayr has ceaselessly pointed out. This did not come easily to early 19th-century naturalists, whose attitudes to the nature of species – classes defined by their essence (or diagnosis) and/or by reference to an archetype (from Aristotle and Plato respectively) – were derived from Linnaeus' methods of classification.

natural selection

At first Darwin's answer to the question, 'under what circumstances would one expect evolution to stop', would have been 'if the environment ceased to change'; i.e. if the extrinsic factors, the ecological *niche* of the organism, remained constant. That would be the case for one species; for evolution as a whole to stop, the whole earth would have to remain unchanging. The most important difference between Darwin's and Lamarck's theory was that in Darwin's theory there was no component of orthogenesis, no innate tendency to evolve. The principal intrinsic factor in an evolving population was its ability to produce, either by chance or perhaps by the 'inheritance of acquired characters', individuals that were slightly better adapted to their environment than were their conspecifics:

> ...if there be, owing to the high geometrical powers of increase of each species, at some age, season, or year, a severe struggle for life...I think it would be a most extraordinary fact if no variation ever had occurred useful to each being's own welfare, in the same way as so many variations have occurred useful to man. But if variations useful to any organic being do occur, assuredly individuals thus characterised will have the best chance of being preserved in the struggle for life; and from the strong principal of inheritance they will tend to produce offspring similarly characterised. This principle of preservation, I have called, for the sake of brevity, Natural Selection.
>
> (*The Origin...* summary of Chapter IV)

At first Darwin saw selection acting only during periods of radical environmental change, so that environmental 'catastrophes' like those reconstructed by Cuvier and Brongniart (pp. 77-8) resulted not in total extinction, but in 'survival of the fittest', to introduce a phrase coined much later by Herbert Spencer (and incorporated by Darwin into the [1869] fifth edition of the *Origin*). By the time of writing of the first edition (1859), however, Darwin had come to see selection as an unremitting force 'daily and hourly scrutinising, throughout the world, every variation, even the slightest; rejecting that which is bad, preserving and adding up all that is good...'. The reason for this constant (metaphorical) scrutiny is the enormous reproductive potential of every member of every species, which is far beyond the reproductive rate needed to maintain population numbers constant:

> Linnaeus has calculated that if an annual plant produced only two seeds – and there is no plant so unproductive as this – and their seedlings next year produced two, and so on, then in twenty years there would be a million plants. The elephant is reckoned to be the slowest breeder of all known animals, and I have taken some pains to estimate its probable minimum rate of natural increase: it will be under the mark to assume that it breeds when thirty years old, and goes on breeding till ninety years old [something of an overestimate of its longevity], bringing forth three pairs of young in this interval; if this be so, at the end of the fifth century there would be alive fifteen million elephants, descended from the first pair.
>
> (*The Origin...* Chapter III)

Wallace, in his paper of 1858 (as sent to Darwin), did similar calculations – 10 million birds from a single pair in fifteen years – but Wallace's presentation of a theory of mechanism differed importantly from Darwin's. Wallace used no metaphor such as 'Natural Selection' to denote his mechanism, but extrapolated from individual cases: 'those [falcons and cats] always survived longest which had the greatest facilities for seizing their prey'. The giraffe did not acquire its long neck by cumulative inheritance of individual stretching (*pace* Lamarck) but because any individuals 'with a longer neck than usual at once secured a fresh range of pasture over the same ground as their shorter-necked companions, and on the first scarcity of food were thereby enabled to outlive them'. Nor did Wallace, then or subsequently, accept the analogy of 'artificial selection' in the breeding of domestic animals developed in the first chapter of the *Origin*, the source of Darwin's metaphor. And Wallace was not a pluralist – with single exception of the origin of the human mind (see p. 19 in this book), all adaptive features of organisms were due to the differential survival to reproduce of *chance* superior variants in populations of organisms, with no 'inheritance of acquired characters'.

But both Darwin and Wallace used the phrase 'struggle for existence' and acknowledged a debt to the Rev. Thomas Malthus' *Essay on the Principle of Population*, first published in 1789 and considerably revised in 1803. As Desmond and Moore have pointed out in their recent biography of Darwin, Malthus' treatise, which concerned human populations, was very much in the public domain in Britain from 1834 onwards, with the passing of the Poor Law Amendment Act. Briefly and brutally Malthus' principle was that any attempt to ameliorate the lot of the lower orders was doomed to failure because they would reproduce to the limit, and beyond, of the available food supply – the poor would always be with us. The only hope was to encourage restraint in reproduction. The spirit of Malthus was embodied in the Act by the Whig government: withdrawal of cash payments to the unemployed, with only punitive workhouses as a hedge against starvation.

the logic of selection

I will set out Darwin's and Wallace's theory as a series of logical propositions. This statement of the theory, taken from my recent *Classification, Evolution and the Nature of Biology*, is modified from an exposition by Ernst Mayr and was further refined in discussion with friends from the former Department of Philosophy, University of Newcastle:

Fact 1: All species have the reproductive potential for exponential growth in population size.

Fact 2: Population numbers normally display stability with limited fluctuations.

Fact 3: Natural resources are limited and, in a constant environment, remain stable.

Inference 1: Since more individuals are produced than can be supported by the available resources, but population size (normally) remains stable, only an (often very small) part of the progeny of each generation survive to reproduce.

Fact 4: There is (phenotypic) variation between individual members of some or all populations.

Fact 5: Some of the variation is heritable.

Fact 6: Some of the variation represents differences in degree of adaptedness to *current environmental conditions.*

Hypothesis 1: Heritable differences in adaptedness occur between members of the same population in some or all populations.

Hypothesis 2: Within a population better adapted individuals have a better chance of surviving to achieve their reproductive potential.

Hypothesis 3: Heritable improved adaptedness (potential or actual) appears spontaneously in individuals within some or all populations.

Inference 2: In the 'struggle for existence' (Inference 1) better adapted individuals increase in frequency in a population with time: *this differential increase is Natural Selection.*

The Theory: Adaptive anagenesis (i.e. adaptive phyletic evolution) is the result of *Natural Selection*, acting on spontaneously occurring adaptive features.

The propositions labelled Fact, 1-6 are all well corroborated. Facts 1-3 and Inference 1 are implicit in what I have said already. The term 'phenotypic' in Fact 4 is one from 20th-century genetics and is explained in chapter 9. For present purposes it can be interpreted as meaning 'detectable but not necessarily heritable'. Fact 4 can be corroborated for our own species by noticing the differences revealed by looking in the mirror as opposed to looking at any other human (unless you happen to be someone's identical twin). Again

Fact 5 is well corroborated, but in Fact 6 I have used the word 'adaptedness' in a special and rigorous sense which I will explain below. Hypothesis 1 is an inference derived from Facts 4-6, but I have labelled it 'hypothesis' because it cannot be deduced logically from those facts, but has empirical content. In other words its truth needs to be tested by observation or experiment. Given Facts 4-6 it *could* be the case that while *some* of the variation within a population of organisms is heritable and *some* of the variation represents differences in adaptedness to the environment, that variation which is inherited does not include any adaptive differences: adaptive differences being purely phenotypic and thus not mirrored in the genetic programs of members of the population.

Hypothesis 2 is also empirical. It *could* be the case that better adapted individuals, falcons and cats which 'had the greatest facilities for seizing their prey' lived exactly as long and produced exactly the same number of offspring as those with the least facilities for seizing their prey, and so on for all features apparently adapting them to their environment – improbable perhaps, but not illogical.

Hypothesis 3 concerns the actual origin of new, better adapted features, or rather individuals with those features. It was assumed *a priori* by Darwin and Wallace. I will discuss our knowledge of how such individuals arise in chapters 9 and 10. Inference 2 then follows from all the previous propositions, *except* Hypothesis 3, and constitutes a *definition* of Natural Selection as Darwin would have understood it. The *Theory* of Natural Selection is a statement of Darwin's and Wallace's theory of evolutionary mechanism.

It will be seen that the theory of Natural Selection is a theory to explain the spread of adaptive features in populations of organisms and, assuming the availability of new adaptive features, the adaptive anagenesis of organisms. There are two important points to make: the theory makes no statement about the origin and mode of inheritance of adaptive features, and, without further propositions does not explain cladogenesis. The popular story is that Darwin, Wallace and their contemporaries knew nothing about genetics, the science of inheritance, but that everything came out all right with the 'rediscovery' of Mendel's pioneering genetic studies in 1900. This is at best a half truth. The phenomenon of genetic dominance was known from work done at the end of the 18th century: Darwin and Wallace knew little of genetics, but not nothing. And then there is the fact that the rapid development of genetics at the beginning of this century produced a reaction away from the theory of natural selection. The reconciliation only came about in the late 1930s and the 1940s.

Both Darwin and Wallace explained cladogenesis in terms of the environment rather than in terms of the genetics of organisms. Darwin's 'principle of divergence' suggested that the variety of environmental resources in any given area would be more fully exploited if a single original species split into a number, each with a unique environmental *niche*, and assumed that selection would bring this about. It also paid him, in trying to convince his readership of the truth of evolution, to downplay any difference in kind between *intra*specific differences (between 'varieties') and *inter*-specific differences (between closely related species). This is echoed by the title of Wallace's 1858 paper – 'On the tendency of varieties to depart indefinitely from the original type'. Modern practice characterises two populations, at least among animals, as distinct species if there are significant barriers to successful inter-breeding between them. I will deal with these matters in the next three chapters. Meanwhile there are three important matters that have led to dangerous criticism of Natural Selection in Darwin's time and since. They are the concepts of adaptation, blending inheritance, and gradualism.

adaptation

One of the most tiresome clichés about the theory of natural selection is that it is a 'tautology'. The people who make this claim do not know what the word 'tautology' means. A tautology is a particular sort of statement that is *logically true*; i.e. is true whatever the subject and predicate of the statement because of the sentence structure. In a tautology the logical truth is determined by virtue of the 'truth-functional terms' it contains. Thus 'it is either raining or it is not' is true because of the presence of 'either' 'or' and 'not'. No one could claim that the theory of natural selection is of this nature. What is being claimed is that it embodies a circular argument along the lines of 'natural selection is the survival of the fittest – how is fitness assessed: by differential survival'.

There are two objections to this claim. In my formulation of the theory Inference 2 incorporates a definition of natural selection. So even if it were true that Inference 2 had no empirical content, the *theory* of natural selection could not be deduced from it. It does not follow logically that a differential increase in better adapted individuals will result in 'varieties [departing] indefinitely from the original type'. The deduction could only be made if evolution were equated merely with increase in the frequency of better adapted individuals within a population. Incredibly, modern population genetics is in danger of making that equation, as we shall see in chapter 12. The second objection to the 'tautology' argument is that 'adaptation' is not equivalent to the enhanced probability of survival to reproduce. The terms 'adapted', 'adaptiveness' (of a trait), 'adaptedness' (of an organism) and 'adaptation' have been clearly defined and decoupled from reproductive propensity by the philosopher of biology Richard Burian:

> Fleetness contributes to the adaptedness of a deer [i.e. shows 'adaptiveness']...if, and only if,...it contributes to the solution of a problem posed to the deer – for example, escaping predation. Fleetness is an adaptation of the deer if, and only if, the deer's fleetness has been moulded by a historical process in which relative fleetness of earlier deer helped shape the fleetness of current deer.

Burian also uses the metaphor of 'engineering fitness' to correspond to 'adaptedness' to emphasise the analogy between adaptation and design.

Originally, of course, the two were equivalent. Adaptations were evidence of design and thus of a designer – evidence for the existence of God. It was surely, at least in part, for this reason that Lamarck felt called upon to explain the origin of adaptations in his theory. The pressure on Darwin to produce a naturalistic theory of adaptation was even greater. There was a long-standing tradition in England of naturalists, particularly clergyman-naturalists to elaborate the 'evidence from design', from John Ray's (1691) *The wisdom of God manifested in the works of Creation*, to William Paley's (1802) *Natural Theology: or evidences of the existence and attributes of the Diety collected from the appearances of Nature.* Darwin and Wallace would also have been aware of Cuvier's opinion (p. 35) that evolution could not take place because change in any attribute of an organism would upset the delicate balance of a working machine – its 'engineering fitness'. They, therefore, saw it as essential to produce a scientific explanation of the origin of adaptations. This preoccupation contributed to the paradox that while the *explanandum* of evolution is the pattern of classification, theories of mechanism from their day to ours do nothing to explain any particular pattern and are weak in explaining cladogenesis, which yields the branching nature of the pattern.

blending inheritance

Many religious people found themselves able to accept that evolution had occurred after the publication of the *Origin* but objected to godless, cruel and wasteful natural selection. But the accepted idea of 'blending inheritance' led to scientific controversy. Just as one might suggest that the inheritance of acquired characters was a commonsense view of part of the cause of genetic change, so, in the 19th century, it seemed to be the sensible view that in sexually-reproducing organisms an offspring would have heritable characters which were an average of those of its parents. So any individual appearing with some feature which better adapted it to its environment would contribute that feature diluted by half to its offspring and there would be successive dilution in subsequent generations.

gradualism

Both Darwin and Wallace insisted that selection acted on tiny adaptive differences between individuals over enormous periods of time – a theory of 'phyletic gradualism' as it is known today. Thus there was a severe problem as to how a new feature yielding a minute selective advantage could make its possessors candidates for selection over many generations before its advantage was diluted beyond the reach of natural selection. Various alternative evolutionary mechanisms were suggested by evolutionists up to the end of the 19th century and beyond – 'acquired characters', divine direction, orthogenesis, evolution by 'saltation' (i.e. in 'jumps') – but the problem was solved (eventually) by 20th-century genetics.

the germ plasm

In this chapter I have principally been concerned with setting out the logic of the theory of natural selection, which is valid as proposed by Darwin and Wallace despite their ignorance of genetics. I will postpone any judgement of the theory until we have seen the modern version developed in the next three chapters. There is, however, one further point to be made at this stage. Towards the end of the 19th century there was a considerable revival of Lamarckism, both the innate tendency to evolve (orthogenesis) and the inheritance of acquired characters, particularly amongst palaeontologists. But belief in the inheritance of acquired characters received a severe blow, not from any observation or experiment (although it has never been satisfactorily demonstrated except perhaps for bacteria), but for theoretical reasons.

These reasons were embodied in the 'doctrine of the continuity of the germ plasm' proposed in 1885 by the biologist August Weismann. In any organism the germ plasm as defined by Weismann is the tissue from which the gametes (sperm or eggs) will develop. The germ plasm, like any other living bodily tissue, is made up of numerous microscopic cells. Weismann was a pioneer in theorising about the material basis of heredity and proposed not only that the genetic program must be embodied in the germ cells, to be passed on to the next generation, but also that it was contained within the nucleus of those cells. But the germ plasm differentiated early in the embryology of the individual parent – how could it be modified by that individual's subsequent history? How could the blacksmith's professional exercise alter the nature of his germ cells in any specific way? Weismann pictured an immortal 'germ line' as it were budding off individual bodies in each generation. In Samuel Butler's sarcastic aphorism 'a hen is an egg's way of producing another egg'.

In biological texts Weismann is also remembered for 'disproving Lamarckism' by cutting off the tails of mice in successive generations and demonstrating that each offspring generation was born neither tailless nor with diminished tails. But Lamarck would have scorned such an experiment – mutilations and the results of accidents were not inherited. What Weismann had disproved (for mice) was part of Darwin's theory of pangenesis, in which 'gemmules', tiny particles, flowed from every part of the body to record events for the next generation by developing as cells in the germ plasm!

notes

As before, I recommend that the best way to understand the theories of mechanism of Lamarck, Wallace and Darwin is to read them in the original (notes to chs 2 and 4 in this book). The development of **Lamarck**'s theory is chronicled in **R.W. Burkhardt**, *The Spirit of System: Lamarck and Evolutionary Biology* (Harvard University Press, 1977). My logical statement of natural selection is modified from that in **Mayr**'s *The Growth of Biological Thought* (notes to ch. 1). **Burian**'s article on 'Adaptation' is in *Dimensions of Darwinism: Themes and Counterthemes in Twentieth-Century Evolutionary Theory* (ed. Marjorie Grene, Cambridge University Press, pp. 287-314). The late 19th-century reaction against Darwinism is described in **Peter Bowler**'s *The Eclipse of Darwinism: Anti-Darwinian Evolution Theories in the Decades around 1900* (Johns Hopkins University Press, 1983).

Orthogenesis: anagenesis (ch. 1) directed by some innate program without reference to enviromental factors: generally discredited.

9: Mendel to Muller

> When we come to the mechanism by which evolution acts, we have made little progress. The mechanism that Darwin himself suggested has been almost universally rejected – in least in the form in which he propounded it. Such broken light as is being thrown on the matter is coming from the study of sciences of which he had never heard.
>
> (Charles Singer: *A Short History of Biology*, 1st edn, 1931)

> [Darwin] had also conquered [at the end of the 19th century] on Natural Selection in the sense that, on a mathematical basis, very small advantages, if inherited, must yield species differentiation. The weak point was in the word *if.* But most evolutionary thought has been deflected in the twentieth century to the nuclear expression of Mendelian principles which are answering the *if.*
>
> (Ibid., 3rd edn, 1959)

inheritance

I mentioned in the last chapter that discovery in 1900 of Mendel's much earlier work on what we now know as genetics produced a reaction against the theory of Natural Selection as the new science developed. The epigraph from the first edition of Singer's history of biology probably represents the majority opinion at the beginning of the 1930s. By the late 1950s the Synthetic Theory, based on Darwin's and Wallace's insights, was (almost) the consensus and Singer had apparently rather unwillingly dragged himself into the second half of the 20th century. The theory of Natural Selection, as set out in our last chapter, is a satisfying logical construction whose empirical propositions are all capable of being corroborated. But in the mid-19th century there was inadequate knowledge of five important fields of biology necessary to give a plausible and fully rounded theory of evolutionary mechanism.

The five fields are:

1. the mode of transmission of hereditary characters from one generation to the next – '*Mendelism*' or more broadly *transmission genetics*;
2. the nature of the hereditary material: how is the genetic program embodied? – *cytogenetics* and *molecular genetics*;
3. the sources of the new variants on which selection is said to act – the study of *mutation*;
4. the genetic changes occurring in interbreeding populations as a result of selection and other factors – *population genetics*;
5. and, most intractable of all, the implementation of the genetic programme in a newly conceived organism to produce the living and developing individual – *developmental genetics*.

In this chapter I shall be concerned with the first three. In chapters 10 and 11 an account of the synthetic theory will involve all five, but with an emphasis on population genetics. I noted in the last chapter the intuitive idea of 'blending inheritance', that any offspring would be, as it were, the genetic average of its parents. But I also said that Darwin and his scientific contemporaries were aware of the phenomenon of what we now call 'genetic *dominance*' – known to Darwin as 'preponderance'. In humans some people cannot taste the bitter principle in grapefruit, the chemical phenylthiocarbamide. Two parents who are 'non-tasters' who have children will produce none but non-tasters, whereas if both parents are 'tasters' (as are about 70% of the population), either may have a hidden 'non-taster' factor. If both do, there is a one-in-four chance of having a non-taster child. As tasting is dominant to non-tasting, so non-tasting is *recessive* to tasting. Again Darwin and his contemporaries were aware of the phenomenon of recessive characters, which appeared to 'skip' one or more generations, but lumped it with 'atavism' – so-called 'throwbacks' to a more primitive state.

Mendel and Mendelism

Johann (Father Gregor) Mendel (1822-84) began his experiments in 1856, three years before the publication of the *Origin*. The principal results were published in 1866. Mendel had returned from completing his education at the University of Vienna, and the experiments were conducted in the monastry garden at Brünn

(Brno), Moravia, where he eventually became abbot. The critical work was on pea plants, *Pisum sativum*. The flowers are normally self-fertilising and Mendel used pea stocks that he knew bred true for a series of characteristics. The characters he chose to study occurred as alternate pairs each in different plants. Thus he could distinguish tall and short plants reared under as nearly as possible identical conditions. Both bred true, but a cross between tall and short plants by artificial cross-pollination yielded only tall plants, whichever parent was the pollen donor i.e. acted as male parent. Tall was dominant to short – subsequently symbolised as *T* and *t* respectively.

Mendel then self-fertilised these 'hybrid' offspring, as had many plant breeders before him, and regained both tall and short plants in the next (F_2) generation. But Mendel differed from his forerunners in that he repeated the crosses many times, for this and the other pairs of contrasted characters that he tested, and always counted the number of offspring of each type. The tall 'hybrids' resulting from the original cross constituted the first filial, or F_1, generation; their offspring by selfing, the F_2.

In the F_2 generation the ratio of tall to short plants was always close to three to one, and this proved to be the dominant/recessive ratio for all other pairs of contrasting characters. Mendel's explanation was that the tall F_1 plants each carried one factor for 'tall' plus one for 'short'. Then their gametes, egg cells or pollen grains, each carried a single factor, with 50% of the egg cells and 50% of the pollen grains carrying 'tall' and 50% of each carrying 'short'. On self-fertilisation of the F_1 plants to give the F_2 generation, each F_2 plant resulted from the fertilisation of an egg cell by a pollen grain. The ratio of the different types of plant resulted from the chances of fusion between the different types of gamete. His prediction was that, with a 3:1 ratio of tall to short plants in the F_2, all the short plants would breed true, but only a third of the tall plants would do so. All the short plants were *homozygous*, having both factors for short, but only a third of the tall plants were *homozygous*, having two 'tall' factors, the rest were *heterozygous*, having one 'tall' and one 'short' factor. He was able to prove this by self-fertilising the tall plants.

We can see why he made his predictions by constructing a simple two-by-two table. The columns represent the two types of pollen grains, the rows the two types of egg cells. The respective products of fertilisation are in the boxes:

Pollen

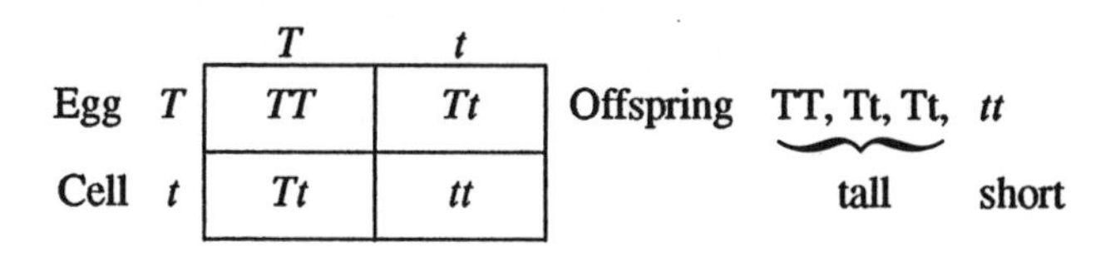

	T	*t*
Egg *T*	*TT*	*Tt*
Cell *t*	*Tt*	*tt*

Offspring TT, Tt, Tt, *tt*
tall short

Mendel then extended his crosses to investigate the behaviour of two pairs of alternate characters in the same plant (a 'dihybrid cross'). His hypothesis was one of '*independent assortment*': each pair of factors would segregate into the gametes quite independently of the other pair. Thus the factor for round seeds in peas is dominant to that for wrinkled seeds, and that for yellow seeds is dominant to that for green seeds. Following the convention of symbolising a pair of factors by the initial of the dominant, we can now construct a four-by-four table to discover the composition of the F_2, with columns and rows headed by each possible type of pollen and egg cell respectively. These will be the gametes from the hybrid F_1 plants all of which will have seeds that appear round and yellow but have a *genotype* (the modern term) of *R r Y y*.

	RY	*Ry*	*rY*	*ry*
RY	*RY* *RY*	*RY* *Ry*	*RY* *rY*	*RY* *ry*
Ry	*RY* *Ry*	*Ry* *Ry*	*Ry* *rY*	*Ry* *ry*
rY	*RY* *rY*	*Ry* *rY*	*rY* *rY*	*rY* *ry*
ry	*RY* *ry*	*Ry* *ry*	*rY* *ry*	*ry* *ry*

The manifestation of the various characters in an organism resulting from the interaction of the genotype with all those factors affecting its growth is the *phenotype* (see ch. 8, p. 95). There are four possible phenotypes in the F_2 of our dihybrid cross, round/yellow, round/green, wrinkled/yellow, and wrinkled/green. Bearing in mind that if a dominant factor is present in an F_2 individual it will determine the appropriate part of the phenotype the reader can discover from the table that the four phenotypes are present in the ratio 9:3:3:1 respectively.

the origin of genetics

Mendel's work was rediscovered in the year 1900 by Hugo de Vries from Amsterdam, the German Carl Correns, and Erich Tschermak from Bohemia. Tschermak was a minor figure but the other two are important to our story. Both had achieved Mendelian ratios in plant breeding, although de Vries did not realise the significance of the 3:1 ratio. Correns, however, did and was the one to phrase 'Mendel's Laws' in their modern form:

1. Of a pair of contrasting characters, only one can be represented in a single gamete [the law of segregation].
2. Each of a pair of contrasting characters may be combined with either of another pair [the law of independent assortment].

Thus, judging from Mendel's results alone, blending inheritance was a fallacy. A recessive character could always reappear in an individual in an animal or plant population, if both that individual's parents were carriers of the factor for that character. New variants arising on a population would retain their integrity for selection to act.

the eclipse of Darwinism

But the pioneers of genetics from 1900 onwards mostly saw the mechanism for evolution in a different light. Evolution occurred by the origin of mutants radically different from the general run of the population and the perpetuation of those mutants by the Mendelian mechanism. There was no 'inheritance of acquired characters' – inheritance was 'hard' in Mayr's terminology – minute differences between individuals, the raw material on which selection acted according to Darwin and Wallace, were of no significance, and selection, if it had a rôle, simply removed grossly non-adapted individuals. A number of 19th-century authors, including Thomas Henry Huxley, had urged that evolution must at times have occurred by 'saltation', but Darwin would have none of it. Saltation became near dogma to the early geneticists of the beginning of the 20th century within the new Mendelian paradigm and influenced particularly by three men: William Bateson (1861-1926), W.L. Johannsen (1857-1927) and de Vries (1848-1935). Bateson, who coined the terms, *genetics, heterozygote* and *homozygote* and also '*allele*' for one of a pair of Mendel's alternate 'factors', was convinced that species differences arose spontaneously without intermediate stages and collected together evidence of discontinuous variation in his massive (1894) *Materials for the study of variation: treated with especial regard to discontinuity in the origin of species*. Johannsen conducted breeding experiments producing a number of 'pure lines' of garden beans. Like Mendel's peas, they are normally self-fertilising, and Johannsen's stocks were homozygous for the characters he was concerned with. But the size of the beans in the pod varied, he concluded, due entirely to environmental factors. The *genes* (his term) for bean size were the same for all plants in a pure line – the plants had the same *genotype* (also his term) for bean size – but the *phenotype* of each bean was different (again his term, but used in the modern sense). So Johannsen concluded that the graded differences which Darwin and Wallace saw as the material for selection were not heritable. Evolution was due to mutation. De Vries claimed actually to have seen the origin of new species by mutation in his breeding studies on the evening primrose *Oenothera lamarckiana* (a somewhat ironical species name!). He certainly produced new variants which bred true but they were not new species, and were due to a complex of genetic causes.

the limits of Mendelism

But while Bateson, Johannsen and de Vries were proclaiming Mendel's contrasting discrete, alternate characters (or rather their causal factors) as the critical differences in evolution, early geneticists were showing that Mendel's Laws were not the universal biological principles they were claimed to be. First, it was not the case that genetic factors always occurred in contrasting pairs, one dominant to the other. Bateson, Saunders and Punnett showed incomplete dominance in 1905: in domestic fowl if black and white chickens of the appropriate breed are crossed, the offspring (the F_1) are 'blue', the Andalusian fowl. When crossed the latter segregate in the ratio 1 black, 2 Adalusian, 1 white, according to Mendelian rules, but neither black nor white is dominant to the other. A similar case in plants is where in *Antirrhinum* ('snapdragon') the heterozygote between red-flowered plants and white-flowered plants has pink flowers.

Nor is it invariably the case that there are only two alternate alleles. An example of this is coat colour in rabbits. The normal 'wild' rabbit pelt is known as 'Agouti', in which each individual hair is grey at the base, then yellow, and outside has a dark tip. Without the yellow band the silvery-grey fur is known as 'Chinchilla', a characteristic breed. Then there is the 'Himalayan' rabbit, which normally has a white coat except that the ears, nose, paws and tail are black. This incidentally demonstrates another principle. When Himalayan rabbits are born, they are all white; the dark areas soon develop. It transpires that this is due to ambient temperature – black pigment develops in hairs growing at a slightly lower temperature in the extremities. If an adult Himalayan rabbit has a small area of white body fur shaved off and an ice pack taped to the bald patch, the new developing hair will be black. Exactly the same phenomena apply to the dark (blue or brown) areas of fur in Siamese cats. In both Himalayan rabbit and cats, hair pigmentation, or lack of it, is controlled by the same allele all over the body, but in interaction with the environment gives two different colours.

Returning to our rabbit series, the last member is the albino with white hair and, unlike all the others, pink eyes. If an albino rabbit is crossed with any of the others, agouti, chinchilla, or Himalayan, it behaves as a Mendelian recessive. But while Himalayan is dominant to albino, it is recessive to agouti and chinchilla. Then chinchilla is dominant to Himalayan and albino, but recessive to agouti, and agouti is dominant to any of the others. Here we have a multiple allelic series, which is also a dominance series (which is not invariably the case).

The early followers of Mendel would like to have believed that each Mendelian factor produced a particular effect independent of other genes apart from its alternate allele(s), but it was soon shown that this was not the case. More crosses by Bateson and Punnett showed this in 1905. In domestic fowl the normal comb on the head is a single fore-and-aft fleshy blade. A shape called 'rose' proved to behave as a simple dominant to 'single'. But a shape called 'pea' was also dominant to 'single'. What then would happen if 'pea' and 'rose' were crossed? Neither was dominant to the other, but the cross produced a new shape 'walnut'. If 'walnut' is crossed with 'walnut', however, for a large number of offspring, the progeny are in the ratio, 9 walnut, 3 rose, 3 pea, 1 single. This is the same as Mendel's dihybrid cross (pp. 105-6); that is there are two pairs of contrasting factors, rose/single and pea/single, so that single can either be absence of rose or absence of pea. Walnut is therefore produced by the interaction of the dominant rose with the dominant pea. This sort of interaction is known as an *epistatic* effect, the interaction of two genes at different *loci*.

Here I introduce the idea that a gene, or better a pair of alleles have a place, a locus, in the genetic code. A pair of genes related as alleles have their own *locus*. The exploration of this fact led to the conclusion that Mendel's second law of independent assortment was false, or rather true only in particular circumstances. *Independent assortment occurred with certainty only when each contrasting pair of alleles, each gene locus, was situated on a different chromosome.*

the material basis of heredity

By the latter half of the 19th century it was established that all complex organisms, animals, plants and many fungi, are built up of *cells*. In most cases these cells are aggregated as tissues, such as (in animals) muscle tissue, nervous tissue, and, more diffuse, blood. Tissues are the building material of organs – individual muscles, the brain, the heart. Tissues grow by multiplication of cells involving division of one cell into two. In most organisms (bacteria, viruses and other primitive forms are an exception) each active cell has a central membrane-bound nucleus, which also has to divide at cell division. In the late 19th century a number of microscopists had shown that at cell division darkly-staining elongate bodies appeared within the nucleus. Whatever the cell type, with one exception, the number of these *chromosomes* was characteristic of the animal or plant. It was eventually agreed, firstly, that while chromosomes were only apparent at cell division they were in fact present in the cell all the time (becoming visible at division due to their shortening and thickening) and, secondly, that they were carriers of hereditary factors. In 1887 Weismann predicted that it would be discovered that *reduction division* occurred in the generation of gametes, the chromosome number being half the characteristic one in each sperm and egg, to be restored to the *diploid* number when they fused at fertilisation. It must have been gratifying to him that reduction division was discovered in the same year. But in the 19th century it was not known whether all the chromosomes in a cell were identical, nor whether each chromosome as a whole was responsible for a part of the organism's body, nor yet whether hereditary factors (in Mendel's sense) were present as single copies or replicated. Many 19th-century biologists (including Darwin) believed in particulate inheritance, but did not have a picture of a genetic *code*.

In 1903 two people, Sutton and Boveri, independently made the specific association between Mendel's factors and chromosomes. In normal (somatic) cells chromosomes occurred in pairs – Mendel's 'factors' occurred in pairs. There were many more factors in any organism than there were chromosomes, so each chromosome had a (probably linear) array of factors (genes). Alternate factors (alleles) occupied corresponding positions on each of a homologous pair of chromosomes (occupied the same locus); and finally, in a cell of any organism, each chromosome of a pair was derived from a different parent and replicated throughout the body.

But if chromosomes were derived half from one parent and half from the other, and the chromosomes retained their integrity from one generation to the next, two different 'factors' (genes) on the same chromosome could not follow Mendel's law of independent assortment, they would be forever *linked.* Ironically linkage was first reported by Correns in 1900. He bred the plants stocks, crossing plants with coloured flowers and 'hoary' leaves with plants with white flowers and smooth leaves. All the F_1 were coloured and hoary; those characters were dominant. In the F_2 Mendelism would have suggested a 9:3:3:1 ratio, but all Correns got was the parental combinations, in fact three coloured/hoary to one white/smooth. Yet white/hoary and coloured/smooth exist, but the flower colour and leaf texture loci are closely linked on the same chromosome. But that was not the whole story, *crossing-over* between loci on pairs of chromosomes also occurs. Bateson, Saunders and Punnett reported a case in 1905. In sweet peas they crossed plants with purple flowers and long pollen grains (both dominant) with plants with red flowers and round pollen (both recessive). As expected all the F_1 were purple and long. With Mendelian independent assortment a 9:3:3:1 ratio would be expected; with complete linkage merely 3 purple/long to 1 red/round. We can summarise their actual results in a table:

	Results	Expected
Purple, Long	284 (74.54%)	56.25%
Purple, Round	21 (5.51%)	18.75%
Red, **Long**	21 (5.51%)	18.75%
Red, Round	55 (14.43%)	6.25%

The expectations represent Mendelian independent assortment, i.e. the 9:3:3:1 ratio. Later Bateson *et al.* corroborated their results with larger numbers of crosses. But if the parental plants had been Purple/Round and Red/Long it would have been those combinations which would significantly have exceeded expectations in the F_2. And furthermore the amount of crossing-over, as measured by the percentage of the two new combinations (purple/round and red/long as in the table) is a measure of the distance between the gene loci, the locus for purple or red and the locus for long or round, on the chromosome.

Drosophila

The convincing link between Mendel's 'factors' and genes on chromosomes was made in the 'Fly Room' at Columbia University, New York, from 1909 onwards. Thomas Hunt Morgan (1866-1945) started to conduct genetic experiments on mammals but soon switched to the fruit fly *Drosophila melanogaster*, one of many *Drosophila* species, most of which feed on the yeasts associated with rotting fruit. *D. melanogaster* has the advantage of very short generation time (about two-and-a-half weeks), toughness and enormous numbers of offspring. It is easily reared on a prepared medium, traditionally in empty milk bottles, and for ease of genetic analysis has only four pairs of chromosomes, which Morgan and his colleagues soon discovered corresponded to four 'linkage groups' established by studying linked Mendelian factors. The chromosomes of a male *Drosophila* as they appear after artificial staining in a cell soon after the beginning of cell division are shown in Figure 9.1. Notice that in the *male* there are three pairs of chromosomes (II-IV) in which each member of the pair is similar to the other, but that in chromosome-pair I, they are dissimilar. An X chromosome is the longer, while a hooked Y chromosome is shorter. In the female there are two X chromosomes and normally no Y. This is also the case in humans and corresponds to the mechanism of sex determination. The system is not universal: in butterflies and birds, females are XY; other animals (and plants) have other mechanisms. But in *Drosophila* there are gene loci on the X chromosome not represented on the Y. The characters produced by genes at those loci, if recessive, are *sex-linked*. Pattern baldness and haemophilia are two human examples. They occur rarely if ever as homozygotes, and thus expressed traits, in the female, whereas the single gene on the single male X chromosome will be expressed.

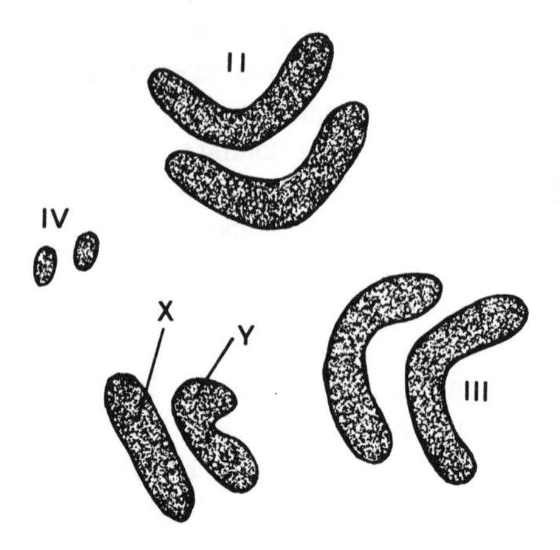

Fig. 9.1 The chromosomes of *Drosophila melanogaster* male as seen in the early stages of cell division.

Morgan's career with *Drosophila* began with the discovery of a new single sex-linked mutant, white eye. Not only did this demonstrate the correspondence between the expression of a sex-linked character and the XY system, but also the fact that the normal gene (red-eye) could mutate spontaneously and then (as he showed subsequently) breed true according to the principles of Mendel modified by linkage. Morgan took on *inter-alia* three brilliant young associates, Alfred H. Sturtevant and Calvin B. Bridges (both originally as undergraduates) and later H.J. Muller (as a graduate student) to form the team of the Fly Room. Sturtevant produced the first *linkage map*, reconstructing the relative position of gene loci along a chromosome by the percentage crossing over between them taken two at a time. Bridges was concerned with measuring linkage and with sex-linkage and discovered double cross-overs. Muller's principal claim to fame was the production and investigation of new mutations.

recombination

In order to understand the physical basis of crossing over we must consider cell division a little further. In normal somatic cell division there has to be a doubling of the chromosome number so that the original number can be restored in the two daughter cells. At cell division each chromosome splits longitudinally into two chromatids, one of which finishes up in each daughter cell as a definitive chromosome. In the case of reduction division there are two cell divisions to produce gametes. In the first division chromatids are formed but in the second the definitive chromosome pairs separate to go to the daughter cells. Crossing over, which in *Drosophila* occurs only in the female (which is by no means always the case in other organisms), occurs as a pair of chromosomes splits to produce four chromatids. The adjacent inner pair of chromatids, one from each chromosome literally 'cross over' (Fig. 9.2), often more than once. Each cross-over point is a *chiasma*, and at the chiasmata the chromatids break and rejoin their opposite number. Thus there is a maternal and paternal contribution to the cells which will further divide to give the gametes.

mutation

Thus by about 1920 the basis of transmission genetics had been established. Segregation and linkage were both explained, and Mendel's hypothetical 'factors' were firmly tied to the concept of linear genes on chromosomes. Later it was shown the genes were not inviolable units, but that crossing over could occur through a gene. So it became necessary to distinguish 'gene' as a unit of function, a unit of recombination, and, later as a unit of mutation. Muller started his studies of mutation in *Drosophila* in the mid-1920s. It transpired that the commonest type of spontaneous mutations were lethals, resulting in the death of the carrier. By using sex-linked lethals as markers and suppressing cross-overs, he was able to detect a new lethal mutant occurring in the sperm of a *Drosophila* male by the death of all its grandsons. The technique is complex and I will not attempt to explain it in detail, but by its use

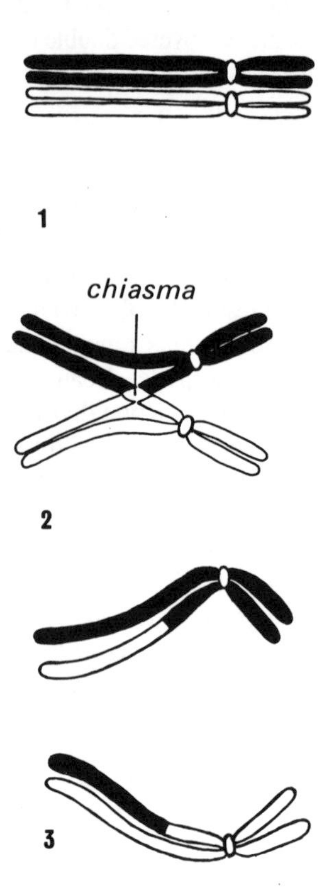

Fig. 9.2 'Crossing over' between adjacent chromatids in *Drosophila*. Black and white indicate each of original chromosome pair. Ovals are centromeres, vital for chromosome movement into daughter cell nuclei.

Muller was able to show that the rate of lethal mutants rose with temperature, and also that X-rays induced mutation. Subsequently it was shown that any ionising radiation, such as gamma rays from radioactive elements, and also ultra-violet light had a similar effect. Many chemicals are also mutagenic: in the 1940s mustard gas and also formaldehyde were shown to be powerful agents.

Two final points must be made. The first is that mutagenic agents increase the rate of mutation, but their effect is in no way directional. One cannot predict which particular genes will be affected. The second point is that, although lethal mutations are common, any gene is liable to mutation (and back-mutation to its original form) and each gene has a characteristic mutation rate without the stimulus of mutagens. Mutations are perhaps the most important, but not the only source of variation on which Natural Selection can act, recombination will also produce new and significant phenotypic characters.

notes

Mendel's work is to be found in most elementary biology texts. A more advanced and scholarly genetics text, in which the matter is arranged historically, is **H.L.K. Whitehouse**, *Towards an Understanding of the Mechanism of Heredity* (3rd edn, E. Arnold, 1973). **William B. Provine** gives details of the early battles between 'Mendelians' and selectionists in his *The Origins of Theoretical Population Genetics* (Chicago University Press, 1971, pbk 1987).

Allele: one of several genes that can occur at the same *locus*, i.e. position on a chromosome, each usually distinguished from others by the contrasting characters that they produce. Corresponds to Mendel's 'Factor'.

Chiasma(ta): breakage points in Chromatids (see below) that allow exchange of lengths of material between maternally and paternally derived chromosomes at cell division – the process of '*crossing over*' resulting in genetic *recombination.*

Chromatids: resulting from the longitudinal splitting of each chromosome into two during cell division. Thus in a *diploid* cell, in which chromosomes occur in (maternal and paternal) pairs, there are four corresponding chromatids. (A complement of unpaired chromosomes in a cell or organism – *haploid*: more than two sets – *polypoid*.)

Dominant/recessive: of the characters coded for by alternate alleles, those that are dominant are manifest in the organism in the heterozygous condition (see below), those that are recessive are not.

Epistatic effect: the production in an organism of a character determined by genes at two or more loci, different from the characters coded individually at each locus.

Gene: variously an allele recognised by its effects (Mendel's 'factor') or a length of DNA recognised as a unit of heredity.

Genotype: originally the total genetic complement of an individual organism (now referred to as its '*Genome*'), now the individuals' genetic constitution (in terms of its alleles) at gene loci under study.

Heterozygous/homozygous: having different/identical, alleles at the same locus on a pair of chromosomes.

(Genetic) linkage: the result of two loci being the same chromosome so that free combination is inhibited. (*Sex-linked* characters are coded for by genes at loci on the sex chromosome.)

Phenotype: manifest characters of an organism resulting from the interaction with the environment during ontogeny (ch. 7) of the genome or the genotype (see above).

10: the Synthetic Theory – genes in populations

the nature of the genome

> We may therefore...regard the great majority of mutant genes, even in man of today, as having a detrimental effect not only homozygously but also heterozygously, insofar as they manifest themselves at all in heterozygotes.
>
> (H.J. Muller, *Our Load of Mutations*, 1950)

> The 'norm' is, thus, neither a single genotype nor a single phenotype. It is not a transcendental constant standing above or beyond the multiform reality. The 'norm' of *Drosophila melanogaster* has as little reality as the 'Type' of *Homo sapiens*.
>
> (Th. Dobzhansky, in *Cold Spring Harbor Symposia* No. 20, 1955)

As the man who had pioneered the study of gene mutation and its induction by radiation, Muller rightly regarded it as his duty to warn mankind of the dangers of radiation of all sorts and particularly, five years after Hiroshima, of nuclear weapons and their testing. But the quotation also reveals something of the attitude of laboratory geneticists to the nature of the genome. (A word is necessary on the term 'genome'. Modern usage is to restrict the word 'genotype' [pp. 105, 116], of an organism, to the state of one or two gene loci under study – heterozygous tall peas have a genotype of *Tt*. But the genome of an organism is its total genetic compliment.) Now geneticists, particularly *Drosophila* geneticists, were concerned with conspicuous and usually deleterious mutants, eyes deficient in pigment, stunted and deformed wings, and especially lethals. Thus arose the concept of the 'mutant' and its normal allele, the 'wild-type'. Muller was concerned with human populations with each member having, by his calculation, between 8 and 80 'mutants' imposing a 'genetic load' on the fitness of the population. We shall return to the concept of genetic load in another context in chapter 12. What I want to emphasise here is a (perhaps exaggerated) picture of the laboratory geneticist's view of the genome from the early years of this century until roughly the time of the Second World War. The normal genome would be homozygous for wild-type alleles at almost all the thousands of gene loci, but a few loci would be heterozygous for, probably deleterious, 'mutants'. How could Darwinian selection act creatively on such a system? Furthermore genetical studies in the laboratory tended to emphasise the individuality of gene loci, each capable of producing its ideal 'wild-type' character.

I have already noted the phenomenon of epistatic interaction between genes for comb shape in fowl (pp. 108-9), but we need to extend the concept of gene interaction in two ways. First epistatic interaction involves, in the simplest case, two genes at different loci producing an unexpected result. There are two other, not clearly separable categories of interaction, modifier genes and polygenic systems. In 1914 Castle and Phillips reported on selection on 'hooded' rats, in which pigmented hair is confined to the head and a stripe down the back. The width of the stripe could be increased or decreased by artificial selection. The 'hooded' gene was recessive relative to uniform colouration, but Castle and Phillips thought that variation in the width of the stripe was due to some sort of variation in the 'hooded' gene itself. Muller and others suggested, correctly, that given homozygous 'hooded' other genes elsewhere on the genome would modify the expression of the character and in 1919 Castle conceded that this was probably the case. Later, as we shall see, Fisher was able to propose that the dominance of a gene could be affected by modifiers.

In 1902 a mathematician G. Udney Yule had suggested that continuously varying characters such as human height could be produced additively by a series of genes at different loci, but in an atmosphere dominated by the views of de Vries and Johannsen he was largely ignored. In fact Yule was right. Another example is human eye colour. At first it may seem that this is simply a case where brown eyes are dominant to blue eyes. Certainly blue-eyed parents normally have exclusively blue-eyed offspring, while two brown-eyed parents may have either brown or blue-eyed children. But there is a spectrum of eye colour from the very rare pink-eyed albino, with no pigment at all in the iris, through blue and various shades to dark brown ('black'). Alleles at several loci are involved and either do or do not contribute to pigmentation. In simplified terms the character is additive: the number of contributing alleles determines the degree of pigmentation.

The other side of the coin from several genes determining one character is several characters determined, or at least affected, by one gene. In rats a mutant causes *achondroplasia*; this, like the similarly named condition in humans, is characterised by the abnormal and deficient development of cartilage, which in humans leads to dwarfism. In rats the mutant produces inability to suckle, faulty blood circulation to the lungs and a restriction of the normally open-rooted and continuously-growing incisor teeth. When a gene affects more than one character its dominance relationships may be different in each case. The mutant *ebony* in *Drosophila* is recessive to the 'wild-type' in causing the characters of a black abdomen (hence the name) and partial blindness, but is dominant to the wild-type in producing hyperactivity including a more vigorous courtship song, produced by rapid vibration of a wing, in the male. In 1978, Kyriacou, Burnet and Connolly at the University of Sheffield were able to show that the action of the ebony gene produced a case of *heterozygote advantage*. Heterozygous males were more successful in courtship, but lacked the homozygous disadvantage of partial blindness.

Thus the relationship of genome to phenotype is a nexus of cause and effect. In fact this is not at all surprising; one can no more atomise an organism into a series of phenotypic characters for genetic analysis than a pheneticist or cladist can atomise the same organism into a series of taxonomic characters for the purpose of classification.

genetics in the wild

By the 1930s there was no evident contradiction between knowledge of genetics and the theory of Natural Selection, but two types of study had to come to maturation before a genetical theory of natural selection could be developed. Both involved studies of genetic variation in whole populations of organisms. The first was a matter of looking at variation in real populations, preferably 'in the field' (as the whole world outside the laboratory is called!). The second was the extension of mathematical genetics from Mendelian ratios to frequencies of alleles etc. in whole populations.

Eminent among students of wild populations was Theodosius Dobzhansky (1900-75) a Russian emigré whose most notable work was done on populations of *Drosophila* in California from the mid-1930s onward. Dobzhansky was reviving a type of research that had been pursued in the (former) Soviet Union in the 1920s, notably by a school of workers led by Sergei Chetverikov. Chetverikov was arrested by Stalin's secret police in 1929 for no stated reason, but population genetics continued in Russia until it (and all other rational genetics) was slowly strangled in the late 1930s and subsequently by Stalin's support for the pseudo-Lamarckian theories of the vicious charlatan Trofim Lysenko. After two years working in Morgan's laboratory, Dobzhansky was based in California from 1929 until 1940, but continued to produce the famous 'GNP' (Genetics of Natural Populations) series of papers with numerous colleagues and assistants until his death in 1975. These papers and other works showed that the 'classical' theory of the genome, as embraced by Muller and other laboratory geneticists was wrong. By breeding wild samples in the lab it was shown that there was enormous variability in wild populations, not just hidden recessives but competing alleles responsive to selection. Furthermore there was extensive *chromosome polymorphism.*

One of the bonuses to geneticists of working on *Drosophila* is that in the salivary glands there are 'giant chromosomes'. Each normal chromosome is represented there by hundreds of copies tightly bound together and of extended length so that each bundle appears as a relatively massive body with a characteristic pattern of transverse stripes like a bar code (Fig. 10.1). Thus gene loci can be assigned a position relative to nearby stripes. In cell division in *Drosophila* and other organisms it is frequently the case that separation of the chromatids after crossing over does not occur cleanly. Bits of chromatid separate in association with the 'wrong' partner chromatid, so that there can be insertions and deletions in the resulting chromosomes. Also lengths of chromosome can become reversed longitudinally – *inversions.* Dobzhansky and his colleagues found that these abnormal chromosomes were retained in wild populations not as occasional freaks but at a high frequency,

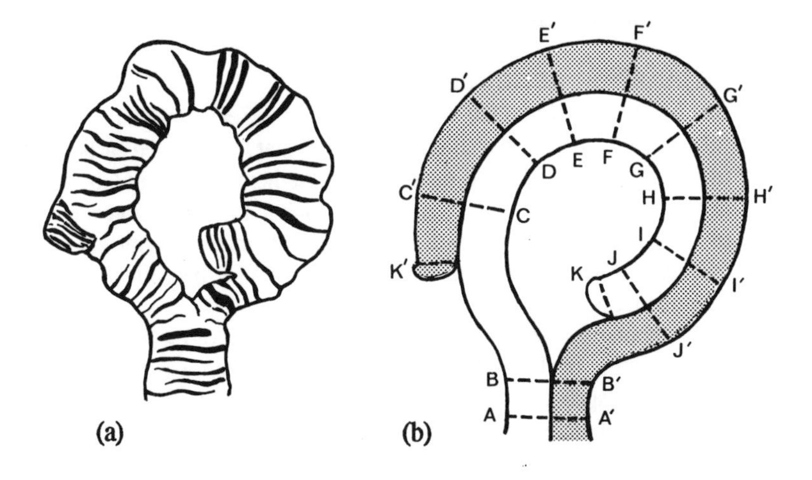

Fig. 10.1 Pairing of the ends of two chromosomes, one with an inversion, in *Drosophila.* (a) As seen in giant chromosome. (b) Note pattern of inversion between loci B' and K' (stippled partner).

often in the heterozygous (or more correctly '*heterokaryotype*') condition. When an 'abnormal' chromosome is paired with a 'normal' partner they go through all sorts of contortions in an 'attempt' to pair corresponding loci, leading to bizarre shapes easily seen in the giant chromosomes (Fig. 10.1).

Hence Dobzhansky showed that the 'balance' not the 'classical' theory of the genome (his words) was correct. Any large population of a species of organism has an enormous reservoir of both patent and hidden variability on which selection can act. This has been corroborated for many other species of animals and plants, notably by the school of English naturalists based at Oxford and led by E.B. Ford (1901-88). Not only is this the case but, as we saw above, specific mutations at a given locus do not occur just once but in many cases are known to occur regularly, however rarely, so that in a large population the same mutation may occur many times in the same generation. So the necessary variation on which Natural Selection can act is there in wild populations. With the phenomenon of inversion, even closer linkage between two gene loci can be selected for, so that, if on the same chromosome, a gene and its modifiers can be tightly linked giving a 'supergene' whose aggregate result is not disrupted by crossing over.

population genetics

But before the 'Synthetic Theory' of Natural Selection could be fully developed the other strand of research that I mentioned earlier in this chapter had to be developed – mathematical population genetics. In order to model the way in which natural selection acts in the wild, population geneticists concern themselves with *gene frequencies* and *genotype frequencies* in whole (if theoretical) interbreeding populations.

In chapter 9 I described the situation at the end of the 19th and the beginning of the 20th century when Darwinism was at a low ebb. Two aspects of the theories of Darwin and Wallace were unpopular: their insistence on gradualism as the normal mode of evolution and natural selection itself. Before the rediscovery of Mendel's work the small differences between individuals on which selection was supposed to act were thought to be progressively cancelled by blending inheritance. So palaeontologists tended to take refuge in 'neo-Lamarckism' and other mystical orthogenetic theories, while students of animal and plant breeding insisted that evolution must proceed by saltation. The latter groups had their views reinforced by Mendelism and the work of Bateson, de Vries and Johannsen (pp. 106-7). But a group of individuals kept faith with both aspects of Darwinism, notably Wallace himself, E.B. Poulton, professor of entomology at Oxford, Weldon and Pearson.

W.F.R. Weldon (1860-1906) a zoologist and Karl Pearson (1857-1936), primarily a mathematician but a remarkable polymath, met as colleagues at University College, London in 1891 and set out consciously to produce a mathematical treatment of heredity, 'biometry', which modelled the inheritance of continously variable characters. This they saw as in opposition to the ideas of Bateson (p. 107) and later to those of all Mendelian geneticists. Yule's suggestion of additive Mendelian factors in 1902 (p. 118) was an attempt at a reconciliation, but the warring factions did not want to be reconciled. When the reconciliation did come it was due to the development of mathematical population genetics, which I can best explain using its parallels with the statistical approach of Mendel's original studies.

the Hardy–Weinberg law

Mendel's triumph was due to the fact that he discovered (or selected!) cases which were simple, with just two alleles, one dominant, one recessive, at a given locus, which segregated normally and independently of those at another given locus. The simplest case in population genetics is similar, but there are other provisos. In the simplest case the population being modelled must be genetically isolated – there must be no migration of individuals to or from other populations or matings with members of other populations, described as 'migration' or 'gene flow'. Members of the population must interbreed freely without restriction or the exercise of choice – the chance of any male mating with any female must be equal: such a (probably non-existent) population is known as '*panmictic*'. Then in the simplest case the model population must be large: in terms of percentages small populations deviate further from statistical expectation. An analogy is tossing a coin. Ten throws are unlikely to yield exactly five heads and five tails, but a thousand 'fair' throws will yield very close to a fifty-fifty result. Finally in our simplest case there shall be no natural selection. With all these unlikely provisos one gets the basic law of population genetics:

> In large panmictic population, with random mating and without mutation, selection, or migration, both genotype frequencies and gene frequencies will remain constant from generation to generation; and the genotype frequencies will be determined by the gene frequencies.

The basis of the law is the so-called Hardy–Weinberg equilibrium. It was pointed out by Castle in 1903 that under the stated circumstances frequencies would remain constant, but the mathematical formulation was derived independently in 1908 by G.H. Hardy, an English mathematician, and Wilhelm Weinberg, a German biologist.

In explaining Mendel's original work I drew a simple two-by-two table to show (p. 105) the results of 'selfing' the (heterozygote) F_1 generation in a 'monohybrid' cross. Representing the allele for 'tall' by T and for 'short' by t we saw that the resulting genotypes in the F_2 were TT, Tt, tt in the ratio 1:2:1. Now while Mendel, or any other geneticist interested in a reliable result, would replicate the cross many times, in essence he was investigating the probable result of crossing two plants in each generation – a 'pure' (homozygous) tall plant with a homozygous short plant from the parental (P) generation, two (or one self fertilising) heterozygous plants in the F_1 generation – to give the characteristic ratio in the F_2 generation. But population geneticists are interested in *gene frequencies*. Given only two alternate alleles at a given locus, say *A* and *a*, the allele or gene frequency is represented by the percentage of *A* or *a* at that locus for the whole population (remembering that each individual organism normally has *two* alleles at the *A*/*a* locus). The convention is that the gene frequency of *A* is represented by **p** and the frequency of *a* is represented by **q**, and both at treated as fractions of the total of all alleles at that locus: thus (importantly) $\mathbf{p} + \mathbf{q} = 1$.

We are now in a position to demonstrate (but not derive or prove: that takes longer) the Hardy–Weinberg equilibrium using our Mendelian two-by-two square but putting in the gene frequencies for the gametes. Multiplying those frequencies in each of the boxes then gives us the *genotype frequencies*, the ratios of the different types of individuals, assuming free interbreeding etc. as in the Hardy–Weinberg Law:

		Pollen (or sperm)	
		p*A*	**q***a*
Egg cells	**p***A*	$\mathbf{p}^2$ *AA*	**pq** *Aa*
	q*a*	**pq** *Aa*	$\mathbf{q}^2$ *aa*

So, given fr. *A* = **p**, fr. *a* = **q**, the genotype frequencies are *AA*: $\mathbf{p}^2$, *Aa*: **2pq**, *aa*: $\mathbf{q}^2$. Notice also that $\mathbf{p}^2 + 2\mathbf{pq} + \mathbf{q}^2$ must equal 1, and also that, by simple algebra, $\mathbf{p}^2 + 2\mathbf{pq} + \mathbf{q}^2 = (\mathbf{p} + \mathbf{q})^2 = 1$. This last is important because it allows us to predict the equilibrium frequency for more than two alleles at the same locus. Thus for three, the genotype frequencies can be calculated from the expansion of $(\mathbf{p} + \mathbf{q} + \mathbf{r})^2$.

origin of the 'Synthetic Theory'

As the rediscovery of Mendel's experiments had formed the basis for transmission genetics, so the insight of Castle, Hardy and Weinberg formed the basis for mathematical population genetics, the modelling of changes in gene frequency and genotype frequency by treating those factors declared constant in the Hardy–Weinberg Law as variables. Three individuals are regarded as the founding fathers of population genetics, R.A. Fisher (1890-1962), originally a statistician, who worked successively at the Rothamsted Experimental Station, University College, London and Cambridge and retired to Adelaide, South Australia; J.B.S. Haldane (1892-1964) also at Cambridge and then University College and finally India; and Sewell Wright (1889-1988) a student of Castle's at Harvard, who then worked for ten years at the United States Department of Agriculture, before moving to the University of Chicago until his (official) retirement, after which he continued his work at the University of Wisconsin. Wright's last paper was written when he had reached the remarkable age of 98.

All three were agreed that natural selection was the principal mechanism for adaptive evolutionary change and all three attempted to model selection in populations of organisms (generally thought of as animals) by assigning 'fitnesses' to, in the simplest case, the three genotypes resulting from two alternate alleles. 'Fitness' was represented in terms of the relative reproductive potential of the three genotypes and conventionally varied between zero and one, but it was an unfortunate term as it did not refer to the adaptiveness of the phenotypic character(s) produced by a particular genotype in a given environment, but was often thought of as an innate feature of that genotype. Given initial gene and/or genotype frequencies, it was then possible by relatively simple mathematics to calculate how many generations it would take for one of a pair of alternate alleles to go to 'fixation' (a gene frequency of one) at the expense of the other, or, alternatively, in the case of (e.g.) heterozygote advantage (p. 119), for equilibrium to be reached. Of course all the other factors in the Hardy–Weinberg Law could be varied – populations with non-random mating, origin and spread of mutants, small populations with 'gene flow' with adjacent populations and so on – in an attempt to simulate mathematically what was happening 'in the field'. Thus a symbiotic relationship among evolutionists arose between population geneticists and practitioners of 'ecological genetics' (Ford's phrase and 1964 book title) or naturalists. Fisher needed Ford as Ford needed Fisher, and Sewall

Wright needed Dobzhansky (whom he helped with the maths of the GNP papers: p. 120) as Dobzhansky needed Wright.

R.A. Fisher and J.B.S. Haldane

Fisher's population genetics was Darwinian in both principal respects: he believed the selection was *the* mechanism of change and he was a gradualist – selection acted to choose between alleles of very small difference in effect which were often additive. He began, in his first important paper in 1918, by partitioning variation in a population for one particular trait (say human height) in terms of *variances*. The total phenotypic variance, a measure of the average deviation of all the members of a population from the mean (height) can be seen as the sum of *environmental variance* and *genetic variance*, assuming additive genes for (height). This led in Fisher's major work, the *Genetical Theory of Natural Selection* (1930), to his 'Fundamental Theorem of Natural Selection': *The rate of increase in fitness of any organism* [i.e. population] *at any time is equal to its genetic variance in fitness at that time*. Fisher regarded this as a law comparable to the physicist's second law of thermodynamics. It has been discussed and rephrased many times, but means in effect that under selection the rate of evolution is proportional to genetic variance. It also produces the irony that as selective evolution proceeds and less fit genotypes are eliminated, it must, *ceteris paribus*, inexorably slow down. Fisher countered this by what is now known at the 'evolutionary arms race'. As a species evolves so do its predators, parasites and competitors – there is thus, from the viewpoint of that species, a continual deterioration of the environment. But Fisher also contended that evolution must take place principally in large populations so that the supply of new mutations, raw material on which selection acted, was sufficient to maintain gradualistic evolution. Haldane agreed with Fisher on large population size, but held that selection coefficients, measures of difference in fitness, were much greater than Fisher allowed, citing the rapid spread of the dark form of the peppered moth in industrial Britain (see ch. 11). Haldane also agreed with Sewall Wright in rejecting Fisher's suggested mechanism for the evolution of dominance.

evolution of dominance

As I have noted before, Fisher believed that dominance had evolved by the selection of modifier genes. He observed that most known 'mutant' alleles were recessive to their respective 'wild-type' alleles at any locus, but suggested that any completely new mutant allele would be neither dominant nor recessive. If it was severely deleterious or lethal in the character(s) it produced, it would be eliminated by selection, but if it were mildly deleterious and particularly if it also produced some advantageous character(s), selection would favour genes elsewhere in the genome which modified the expression of the mutant so that it became recessive, or more correctly its deleterious characters became recessive to those of the 'wild-type'.

Both Haldane and Wright objected to this scenario. 'Modifier genes' would be selected for their own effects rather than for the modification of very rare mutant alleles, and in any case, the evolution of dominance would be so slow that the rare new mutant would disappear by selection or chance before modification could occur. Fisher's rejoinder was that if the mutant and its wild type allele showed heterozygote advantage for any character, the mutant would be held in the population long enough for modification to occur. It is certainly true that *artificial* selection for dominance would work. In 1940 Ford showed that either of a pair of alternate alleles could be made dominant in the magpie moth (Fig. 10.2). In this moth, whose caterpillars are a pest on currant bushes, the normal form has a white background colour to the wings, but there is a yellow form (*lutea*) produced at the same locus. Heterozygotes

Fig. 10.2 The magpie moth, *Abraxas grossulariata* – stippled areas are bright orange.

are normally intermediate in colour, but by successively breeding from the yellowest heterozygotes in each generation he got heterozygotes well within the homozygous *lutea* range in only four generations. Making white dominant was even quicker. By breeding from the whitest heterozygotes, he was able to make white dominant in only three generations.

But Haldane and Wright rejected Ford's experiments as irrelevant. Such rigorous selection would never occur in the wild on Fisher's scenario – it was the slow pace of the latter which made it improbable. Needless to say, both Haldane and Wright had their own theories, which were similar but not identical. They both pointed out that many recessive mutants, such as 'white eye' in *Drosophila*, were identified with characters caused by some biochemical deficiency, such as lack of an enzyme in a biochemical pathway. If the dominant allele then produced twice the amount of enzyme required when homozygous, that would not matter – enzymes are catalysts in biochemical reactions, a surplus does not affect the result. But in the heterozygote there would be 'enough for two' to mask the deficiency of the mutant. Selection for this situation (Haldane suggested 'gene doubling': two adjacent loci) would make the wild type dominant.

Which theory is correct? Probably both – if all new mutants are merely distinguished by deficiency, it is difficult to see how evolution could advance, but some are so distinguished. Fisher's scenario, on the other hand, may well apply to mutants which contribute significantly to the raw material for creative selection to work on. But it is not only mutation of genes that produces heritable phenotypic characters on which selection can act, recombination as a result of crossing over (pp. 110-11, 113-14) and chromosome mutations would produce new interactive gene combinations and thus new characters.

Wright and the 'shifting balance' theory

Sewall Wright's theory of the evolution of dominance may not be the whole story, but his view of evolution within species is almost certainly more realistic than that of Fisher or Haldane. Their emphasis was on evolution in large populations in which the same mutant would crop up many times in the same generation. In Wrights '*shifting balance*' theory, the membership of a species is grouped as large numbers of small populations or *demes* with varying amounts of gene flow, as migration or outbreeding, between them. His theory is much more in terms of adaptive change in these *demes*, rather than of the fate of particular genotypes in large populations. To understand it two concepts must be grasped: the metaphor of the '*adaptive landscape*' and the phenomenon of '*genetic drift*'. The adaptive landscape is symbolised by a map in which there are peaks and valleys defined by contour lines. A deme can be pictured as moving through this metaphorical landscape in evolutionary time. Natural Selection will increase the net fitness of the deme, so that it will ascend an adaptive peak representing a local optimum of adaptive genotypes. But the peak attained by a deme depends on its previous position in the landscape – it may not be the highest peak, i.e. the best possible degree of adaptation available to the gene pool (total genetic compliment) of the deme.

Wright suggests that genetic drift may allow a deme to move down the 'slope' of one peak, eventually to ascend a higher one, against the direction of natural selection. One component of genetic drift is sampling error. On p. 123 I pointed out that a small sample of (e.g.) heads and tails in throws of a coin may deviate quite widely from the statistical expectation of fifty-fifty. This type of deviation will occur with Mendelian ratios and expected gene frequencies in populations if the sample is small. But it will also occur against the expected results of natural selection, as measured (or modelled) in terms of the relative fitness of two competing genotypes. In a *very* small deme this would result in the random fixation and loss of genes impoverishing the gene pool; but in a somewhat larger deme, slightly disadvantageous genes or gene combinations will fail to be eliminated by selection, hidden recessives will be exposed as homozygotes to produce new phenotypes, and new epistatic effects will be set up. All this non-selective change is represented by descent from a minor adaptive peak, and it is enhanced by gene flow with (literally) adjacent demes. Selection can then act to move the deme up the slope of a higher peak, meaning that the deme, and the species it represents, is more highly adapted to its ecological niche.

The work of Fisher, Haldane and Wright can be compared, in the history of the Synthetic Theory of evolution, to enabling legislation. It convinced geneticists, taxonomists, field naturalists and palaeontologists that Mendelian genetics was not incompatible with the heterogeneity and gradation they saw in organisms in the field and with the nature of the fossil record. The synthetic theory followed, manifested particularly in a few very influential books published between 1937 and the end of the Second World War. We begin our next chapter with these.

notes

Provine's (1971) history (notes to ch. 9) gives an account of the early history of population genetics and the work of **Fisher, Haldane** and **Wright** up to the early 1930s. His more recent and ambitious biography of Wright, *Sewall Wright and Evolutionary Biology* (Chicago University Press, 1986), apart from an account of Wright's life and career, which spanned nearly 80 years, includes discussion of the development of evolutionary genetics through this century and is particularly good on Wright's interaction with Dobzhansky and Fisher. The characteristically English approach to evolutionary genetics is examplified in **E.B. Ford**'s *Ecological Genetics* (4th edn, Chapman & Hall, 1975). **Dobzhansky**'s GNP papers (in which Wright was heavily involved) are collected together in R.C. Lewontin *et al., Dobzhansky's Genetics of Natural Populations I-XLIII* (Columbia University Press, 1981) with a biography of Dobzhansky by Provine and an appraisal of the papers by Lewontin. The example of heterozygote advantage in *Drosophila* is in **C.P. Kyriacou, B. Burnet** and **K. Connolly**, *Animal Behaviour* 26 (1978) 1195-206.

Deme: a naturally-occurring, partially isolated population of a particular species. The genetic complement of the whole deme is its '*gene-pool*'; any outbreeding with members of other demes is '*gene-flow*'. If interbreeding between individuals within the deme is random, the deme is '*panmictic*'.

Fixation: over a number of generations, when of two or more alleles at the same locus in the gene pool of a population all are eliminated (by selection or genetic drift: see below) except one, that one will then have a '*gene frequency*' of 100% or will have gone to fixation.

Genetic drift: change in the gene frequency of a particular allele over the generations (even resulting in elimination or fixation) due to random statistical effects rather than natural selection.

Hetero-/Homokaryotype: corresponding to hetero/homozygote

(ch. 9) but for whole chromosomes. Irregular crossing-over and/or *inversions* of parts of chromosomes will result in some of the gene loci of one chromosome not lining up with those of its partner. This is the heterokaryotype condition.

'*Wild type*': used by laboratory geneticists to refer to the 'normal' allele at the same locus as a conspicuous mutant, or to the 'normal' organism.

11: the Synthetic Theory – speciation and polymorphism

the books

> The time is ripe for a rapid advance in our understanding of evolution. Genetics, developmental physiology, ecology, systematics, palaeontology, cytology, mathematical analysis, have all provided new facts or new tools of research: the need today is for concerted attack and synthesis. If this book contributes to such a synthetic point of view, I shall be well content.
>
> (Julian Huxley, *Evolution the Modern Synthesis*, 1942; Preface)

The 'Synthetic Theory of Evolution' takes its name from the title of Julian Huxley's book. It is often conflated with 'Neo-Darwinism', but the latter term was coined by Darwin's former protegé Romanes in 1896 to characterise a version of the theory of Natural Selection in which, following Weismann, the Lamarckian 'inheritance of acquired characters' was rejected. But Huxley's book was not the first of the recognised texts of the Synthesis, that was Dobzhansky's (1937) *Genetics and the Origin of Species*, written before he embarked on the 'GNP' papers (p. 120).

In his book Dobzhansky emphasises genetic variation in the wild, including geographical variation, and gives an account of the mechanism of evolutionary change that draws heavily on Wright's shifting balance theory. But Dobzhansky goes on to consider problems of speciation, the splitting of one species into two or more. It is then clear that, at least in animals, 'the origin of species' is the origin of 'isolating mechanisms', those factors that prevent or inhibit reproduction between members of two distinct but closely related species. The factors can prevent mating between members of the two species or reduce or eliminate the viability or fertility of the offspring of the mating.

The second great book of the synthesis was Ernst Mayr's (1942) *Systematics and the Origin of Species*. He too was much pre-occupied with speciation. Mayr's research had been principally in ornithology, particularly the classification and general systematics of birds in New Guinea and the neighbouring areas of the 'Malay Archipelago', as had that of Alfred Russel Wallace. Thus, as one might expect, Mayr's approach to the then current problems of evolution was through taxonomy. The book had a major and a minor theme – variation within and between species, leading to development of a theory of speciation, and development of a natural classification. Central to both themes was his attempt to elucidate a 'biological species concept' (see below). Subsequently each theme was developed in important books – Mayr's (1963) *Animal Species and Evolution*, and a succession of books on the principles of classification.

Huxley's book was perhaps wider-ranging in its aims but lacking in the intensity of focus of Dobzhansky's or Mayr's. He attempted a review of all aspects of evolutionary theory as it was understood at the time, incorporating his own particular interests in natural history, especially the behaviour of birds, and allometric growth. A fourth founding father of the Synthetic Theory is usually recognised – the American palaeontologist, George Gaylord Simpson. Simpson's (1944) book, *Tempo and Mode in Evolution*, was in part devoted to demonstrating that the fossil record was not imcompatible with the emerging synthesis, and also to propounding the author's distinctive ideas on the pattern and processes of phylogeny. I will refer to these ideas in chapter 12.

speciation

In their pioneering work of the development of population genetics, Fisher, Haldane and Wright had not devoted much time to modelling speciation. The assumption was, and in some circles still is, that the processes of natural selection could be extrapolated to account for change within populations, anagenetic change (phyletic evolution), cladogenesis (speciation) and all the large scale phenomena of evolution as well – all the 'modes' of evolution, to use Simpson's word. Darwin had assumed that the 'origin of species' resulted simply from evolutionary change with time. A population would become sufficiently differentiated from its ancestral stock to achieve the status of a new species. It would also have diverged sufficiently from another population descended from the same stock by being selected for a different ecological niche for the two to have achieved specific distinction. Darwin did not see speciation as an event. It is to the eternal credit of Dobzhansky and Mayr particularly that they did so (and to that of Hennig that he incorporated that insight into taxonomy: ch. 5).

species concepts

A necessary prerequisite for producing generalisations about speciation was a '*biological species concept*'. There are two components of this. The first of these is that species, or rather their members, breed true. This was enunciated by John Ray in 1686 in his *Historia Plantarum.* The second component is embodied by Buffon in the fourth volume of his *Histoire Naturelle* (1749-1804), the idea of a species as a group of individuals that can interbreed amonst themselves, but are incapable of producing fertile offspring by breeding with members of other species. Buffon cited the horse and the ass, whose offspring the mule is invariably sterile. In his first volume, however, Buffon proposed what Mayr was to call a 'typological species concept' based on joint possession of unique characters.

Despite the insights of Ray and Buffon and many naturalists since, the biological species concept was not generally accepted until the synthesis, or rather did not penetrate the consciousness of taxonomists. Even then there were problems of definition. Mayr's 1942 formulation was: '*Species are groups of actually or potentially interbreeding natural populations which are reproductively isolated from other such groups*'. Later he was to remove the phrase 'actually or potentially'. His latest effort (1982) is: '*A species is a reproductive community of populations (reproductively isolated from others) that occupies a specific niche in nature*'. Present day objections to the latter definition are (1) that it implies an 'isolation species concept', whereas what is required is a 'recognition species concept' emphasising the recognition of a member of a species by a potential mate; (2) that 'occupies a specific niche in nature' imports Aristolelian essentialism ('typology') by the back door – it is just as bad to define a species by a unique series of ecological characteristics as by a unique series of anatomical characteristics; (3) that the definition is zoocentric – quite other considerations apply in plants; and (4) that the definition is not even potentially useable in dealing with fossils.

The first objection is easily dealt with. It has been pointed out that the isolation and recognition concepts are simply opposite sides of the same coin. The second objection may be regarded as resulting from Mayr's attempt to take account of the third. Evolutionary theory from the development of the synthesis to the present day has been elaborated almost entirely by zoologists (including palaeozoologists). The situation in plants with regard to speciation is very different from that in animals. Many plant species interbreed quite freely with other closely related plant species – plant species *are* defined in terms of appearance and ecological niche. The corresponding species concept is a 'cohesion concept' – '*a species is the most inclusive population of individuals having the potential for phenotypic cohesion through intrinsic cohesion mechanisms*' (Templeton, 1989). According to Templeton the ability to interbreed is one cohesion mechanism, but it is neither necessary nor sufficient to delimit a species, the other principal cohesion mechanism is 'demographic exchangeability'. All the members of a species occupy the same environmental niche and in any given area an advantageous mutation will go to fixation, i.e. its possessors and their descendents will have, and realise, the potential eventually to comprise the whole population. Thus individual organisms (notably plants and asexual organisms such as bacteria) can compete to the death with one another without any necessary exchange of genetic materials in sexual reproduction.

isolating mechanisms

I will return briefly to plant speciation, but first let us look at the 'Synthesis model' of the origin of isolating mechanisms. Mayr divided isolating mechanisms into postmating and premating mechanisms (although the word 'mechanism' in this and other biological contexts implies teleology – final causes – in a way which I find objectionable). Postmating mechanisms result in the reduction of viability or fertility of the hybrid offspring of two different species, or, *in extremis*, failure of fertilisation. In some 'sibling-species' matings in *Drosophila* the female produces different types of antigenic reaction against 'foreign' sperm – the sperm actually break up (lysis), there is a swelling of the vagina which impedes or kills them, or the sperm cannot penetrate the egg membrane. Even if fertilisation is achieved the embryo may fail to develop, or the offspring may have lowered vitality or fertility. Dobzhansky and his colleagues in their GNP work and before realised that two 'races' of *Drosophila* on which they worked in the field were in fact separate *sibling species* (Mayr's term), closely related, difficult to distinguish and with broadly overlapping ranges. In an interspecies cross between the two (*D. pseudoobscura* and *D. persimilis*) the F_1 females are fertile, but the F_1 males are usually sterile; incidentally illustrating one of the few 'laws' of speciation, 'Haldane's Rule', that in such crosses it is the offspring belonging to the heterogametic sex (XY: see p. 111-12) which are sterile. Another example of conflated sibling species was the European malarial mosquito '*Anopheles maculipennis*' which puzzled medical entomologists in the early years of this century because of the different biology and habits of different populations. It subsequently transpired that six species had been conflated. Of these the males of one would mate with females of four other species, but following Haldane's rule all F_1 males were sterile. F_1 females were variously fertile (2 spp.), sterile (1 sp.) or had a genome which was lethal (1 sp.).

Premating mechanisms need not imply any genetic incompatibility between two species. Two British species of duck, the Mallard (*Anas platyrhynchos*) and the Pintail (*A. acuta*) interbreed freely in captivity; in fact, several duck species will live in happy interspecific promiscuity in wildfowl refuges and produce fully fertile hybrids, but rarely if ever cross in the wild. Both species have males which acquire a striking species-specific plumage in the mating season, but for the rest of the year go into 'eclipse', moulting to the same drab brownish camouflage as the females. The fact that the male's courtship plumage is a matter of species recognition – the female will only mate with 'correct' males – is corroborated by sites where only one species occurs. Mallard but no Pintail occur on the Hawaiian islands, Pintail but no Mallard occur on the Kerguelen islands, far south in the Indian Ocean. In both cases males are in permanent eclipse. Species descrimination is unnecessary: camouflage has been (naturally) selected.

Other examples of premating mechanisms involve species recognition, visual in (e.g.) fireflies in the Caribbean each with a characteristic colour and/or pulse of light flash; auditory as in birds, frogs and grasshoppers; or chemical (scent) as, notably, in many species of moth, where the sedentary females produce characteristic pheromones and the males can detect just a few wind-borne molecules. But cases where potential mates cannot mate for mechanical reasons, or do not meet despite being *sympatric* (having overlapping species ranges) because of differences in mating time, seasonal (e.g. the British Herring Gull and Lesser Black-backed Gull) or diurnal (preferred mating hour, e.g. sibling *Drosophila* species) are also regarded as examples of premating mechanisms.

geography and speciation

Thus in sexually-reproducing animals, the 'origin of species' is 'the origin of isolating mechanisms'. Throughout a long career Mayr has striven, with some success, to argue that such speciation is always *allopatric*. Allopatric populations are those separated from one another by some barrier wider (or more obstructive) than the normal 'cruising range' of either. According to Mayr there must be an allopatric phase between two populations in the process of speciation before they acquire isolating mechanisms separating one from the other. There are two principal ways in which a single population of animals could become divided geographically: *vicariance* and *dispersal*. A vicariance event is the origin of a barrier which splits a pre-existing population (ch. 7), while dispersal implies a sample of population crossing a pre-existing barrier. Various scenarios are possible for the latter. 'Voluntary' dispersal involves such scenarios as floating vegetation mats, mud on birds' feet etc. Related to dispersal is Mayr's concept of the *founder principle*. A very small sample of a large population, down to a single gravid female, which reaches a new environment, such as an oceanic island, will represent an impoverished and atypical sample of the gene pool of the parent population. Its descendents are thus given a 'kick-start' to differentiation and speciation, particularly with rapid population expansion. Wright's genetic drift and shifting balance will play an important part.

selection and speciation

If we assume, as I believe to be the case, that Mayr is correct is saying that in sexual animals most or all speciation involves an allopatric phase, then one can ask, 'what part, if any, does natural selection play in speciation?' This question can be posed as two subordinate questions. The first is to ask whether two newly-evolved sister species can achieve genetic isolation without becoming secondarily sympatric (or *parapatric*: see below). The answer is 'yes'. A striking case is the population of *Drosophila pseudoobscura* (Dobzhansky's favourite species!) from Bogatá, Columbia, South America. The normal range of the species extends from the western USA down into Mexico and Guatamala: the nearest known population from Bogatá is 1500 miles away in Guatamala. In a survey in 1955-6 no *D. pseudoobscura* were found amongst numerous *Drosophila* species in Bogatá. The species was first found there in 1960. By 1962 about 50% of all *Drosophila* in some areas were *D. pseudoobscura*. In 1967 experimental crosses were tried between Bogatá and the main American stock: if Bogatá females were crossed with main stock males, all the F_1 males were sterile (Haldane's Rule), but the reciprocal cross (Bogatá males x main stock females) gave fully fertile offspring of both sexes. Nevertheless the degree of genetic isolation is sufficient to accord Bogatá species status. A new species appears to have evolved in less than ten years. No doubt selection for the new environment and genetic drift both contributed, but there could not have been selection for either premating or postmating mechanisms.

The second question is to ask what types of selection enhance such mechanisms when there is no geographical barrier.

There are in theory three types of such selection: (1) against hybrids which are less viable or fertile, enhancing postmating isolation; (2) reduction of hybrid production by evolving 'positive assortative mating' (like mates preferentially with like) – known as 'reinforcement', or the 'Wallace effect' after Alfred Russel Wallace; (3) 'reproductive character displacement': selection for increasing the distinction of species (or incipient species) recognition, enhancing premating isolation.

hybrid zones

In most cases it is difficult to find cases of such selection in pairs of sympatric species – it has all happened, so attention has focused on *parapatric* distributions where the two populations are adjacent without an evident barrier, but with little or no overlap. In almost every case of pairs of 'semi-species' which are thus adjacent it seems certain that they were once allopatric, often in *refugia* during the recent Ice Ages, and have expanded to meet, often to produce a 'hybrid zone'. Some zones include specimens of the two parent species, including mosaic zones in which the environment is patchy favouring one or the other. Of greater interest are 'tension zones' in which only hybrids are present, the tension being between selection against the hybrids, causing the zone to narrow, and the lifetime movement (vagility) of the hybrids themselves causing the zone to expand with introgression each side into the parent populations. The former leads to speciation, the latter to merging of the two species.

An example of such a tension zone is between two sister species of European toad, the fire-bellied toad and the yellow-bellied toad – both are brightly-coloured underneath and roll over to expose the warning colour, backed up by poison glands in the skin, when threatened. The former toad is a lowland form occurring in the east, the latter a mountain form in the west. The hybrid zone occurs in eastern-facing foothills. Genetics of the hybrids in the zone has been studied by N.H. Barton and colleagues. They looked at five gene loci coding for enzymes, with each parent species having a different characteristic gene at each locus. Thus outside the hybrid zone the frequency of the five 'fire-bellied' genes was each 100% to the east and that of the five alternate 'yellow-bellied' genes was 100% in the west. In the zone there was a steep *cline*, a gradient of frequency from one state to the other for all five loci, suggesting strong selection against the hybrid, heterozygous state – two parapatric populations in secondary contact and well on the way to speciation.

So selection against hybrids is probably important when incipient species, after an allopatric phase, become parapatric or sympatric, but there is disagreement about the existence of the Wallace effect. On the other hand reproductive character displacement is known in two species of Australian frog whose male mating calls are almost identical in allopatry but diverge from one another in a broad area of overlap in the state of Victoria.

sympatric speciation?

But the sixty-four thousand dollar question, still debated vigorously, is whether *sympatric speciation*, without an allopatric phase, can occur. For plants the short answer is 'yes'. Hybrid species must have arisen in sympatry with their two parent species. If a hybrid plant undergoes polyploidy (doubling up of the chromosome number because of abnormal cell division to give a tetraploid), then it will be reproductively isolated from its parent species. Self-fertilisation or vegetative reproduction will then allow its descendents to continue as a new species. It is debatable when this happens in animals, but other scenarios are suggested in which assortative mating occurs among insects which have taken spontaneously to a new foodplant.

At the time that the synthetic theory was getting under way a strongly dissenting voice made itself heard. Richard Goldschmidt, a German refugee in California and an eminent geneticist, published his *The Material Basis of Evolution* in 1940. To him speciation was an entirely different process from the accumulation and fixation of numerous one locus mutations as modelled in population genetics. Every so often there would be a massive reorganisation of the genome of an individual organism resulting in a 'monster'. Most of these would be inviable, but, in his most famous phrase, the occasional 'hopeful monster' would be pre-adapted to a new environmental niche and thus become a new species. Macroevolution, initiated by the origin of species, was decoupled from microevolution, all changes within species.

mimicry

As the synthetic theory 'hardened' (Gould's word) after the Second World War, Goldschmidt came to be seen as a heretic by the orthodox, to be refuted at all cost. One chosen battle-ground between Goldschmidt and the selectionists was the phenomenon of butterfly mimicry. This had been of interest to evolutionary theorists ever since Henry Walter Bates, companion to Wallace in Amazonia (ch. 2), returned to England in 1859. Bates had seen a number of cases where common and, as he surmised, poisonous species of butterfly, with conspicuous warning colours, had unrelated and much rarer species, surmised to be palatable, associated with them and closely similar in appearance. This he subsequently explained (in 1862) as a case of natural selection of the palatable species to look like the poisonous one and thus repel predators such as birds. It was the first use of the theory of natural selection by someone other than Darwin or Wallace. Later Wallace was to interest himself in *mimetic polymorphism* in a butterfly of the large swallowtail family, the species *Papilio polytes*.

Genetic polymorphism is defined by E.B. Ford:

> ...the occurrence together in the same habitat at the same time of two or more forms of a species in such proportions that the rarest of them cannot be maintained merely by recurrent mutation.

Mimetic polymorphism is a special case of genetic polymorphism in which at least some of the 'morphs' are mimics of different poisonous models. Usually, but not always, in such polymorphic species the males all look alike and are non-mimetic (allowing species recognition by the females), but the females are polymorphic. This is the case in *Papilio polytes* and also in the African species *Papilio dardanus*. In the former the females mimic several species of not closely related poisonous swallowtails, in the latter they mimic poisonous butterflies of two taxonomically distant families.

Goldschmidt believed, as had the early Mendelians, that mimetic patterns had arisen in one go by the same sort of macromutation that resulted in instant speciation. Supporters of the Synthesis, following Fisher and Ford before the Second World War, accepted a Darwinian scenario of small changes, gradualism, mediated by Natural Selection. Soon after the war, in the 1950s mimetic polymorphism, particularly in *Papilio dardanus*, became a test case. Philip Sheppard (formerly one of the Oxford group) and Cyril Clarke, both at Liverpool University, began a brilliant series of genetic experiments with different morphs of the butterfly from different parts of Africa. Although all the males look more or less alike and are non-mimetic, each has the genotype of a particular morph which is manifest in the corresponding female. The characters seen in the phenotype are *sex-limited* (Fig. 11.1). Also, and importantly, the different mimetic morphs, or at least the first three of the four principal ones shown in Figure 11.1 form a dominance series (see p. 108). One further fact is necessary: *P. dardanus* occurs on the island of Madagascar as well as in most non-desert areas of the African mainland, but on Madagascar all the females are similar, non-mimetic, and look like the males.

Clarke and Sheppard reasoned that if the different female patterns had each arisen in one go by a macromutation then no plan of cross-breeding could disrupt them, but if they had arisen as a series of mutations retained by selection, suitable crosses should reveal the different elements of the pattern produced by different genes.

Fig. 11.1 The mimetic butterfly *Papilio dardanus* and its models. Non-mimetic male, top centre; females, right, opposite their poisonous models, left. Fine stipple, lemon yellow; medium stipple, dull yellow; dark stipple, orange.

They further reasoned that if dominance had evolved by the selection of a series of modifier genes, there would be no dominance in a population where the mimetic morphs did not exist. Their selectionist views were vindicated by crossing specimens from the main African stock with specimens from Madagascar. Female hybrids were intermediate between the mimetic morph and the male-like-females and highly variable. Dominance modifiers did not work outside mimetic populations. Clarke and Sheppard were also able to show that the wing patterns were controlled by a series of genes, not a single mutant. They also got similar results with *Papilio polytes* and another Asian polymorphic mimic *P. memnon*.

selection and polymorphism

The explanation of genetic polymorphism in terms of natural selection was regarded as a particular triumph of the Synthetic Theory. Another example was that of the banded snail, *Cepaea nemoralis*, and its sister species *C. hortensis*. Before leaving Oxford, Philip Sheppard had investigated this with his colleague Arthur Cain. Both species of snail have extensive visible polymorphisms which appear to be homologous and were thus probably present in their common ancestor. The shell can be yellow, pink or brown, with variants on each colour. The shell bands spiral parallel to one another from the apex to the mouth of the shell. The normal maximum is five bands, which are usually dark brown, but the number may vary from zero to five, with zero, five and one band in the number three position ('midbanded') common. But the bands may be wide ('spread bands') and even run into one another laterally, obscuring the background colour. There are also rarer morphs for atypical band colour and continuity – and then the colour of the animal within the shell varies.

Cain and Sheppard set out to show, from extensive population samples in the Oxford area, that shell colour and pattern were adaptive as camouflage. There would be many types of background on which the snails lived within the area occupied by a single population. The authors showed a significant correlation between high frequency of pink and brown shells and woodland backgrounds with leaf litter, and between high frequency of yellow shells and green backgrounds – hedgerows and open fields. Furthermore variegated backgrounds, hedgerows and coarse vegetation had significantly more banded shells than the more uniform cropped or mown turf. But the story got more complicated, which is often the case when biological generalisations are attempted. In other parts of Europe temperature seemed to be involved in morph frequency and even in Britain there were populations showing 'area effects', notably on the Marlborough Downs – adjacent populations on similar backgrounds with strongly different morph frequencies, each perhaps derived from small 'founder' populations.

The type of genetic polymorphism seen in mimetic butterflies and *Cepaea* snails is a *stable* polymorphism. In fact there are subfossil population samples of *Cepaea* with not only the same morphs as the extant population in the same area, but similar frequencies. Another type, although not clearly distinguished, is *transient* polymorphism in which one of two or more alternate characters is on its way to fixation or extinction. The best known case for all, at least British, school and university students is that of the peppered moth, *Biston betularia.* This is presented in school texts as *the* example of Natural Selection.

'industrial melanism'

In unpolluted woodlands the '*typica*' form of the night-flying moth is camouflaged as it rests during the day on trees covered with greyish lichens, but in the middle of the last century a dark melanic, almost black, form of the moth was recorded in Manchester, apparently camouflaged against soot-covered, lichen-free trees. Today the melanic form (*carbonaria*) is completely genetically dominant to *typica*, but there are some intermediate forms, collectively known as *insularia.* All three named forms appear to be produced by a series of alleles at the same gene locus. In the 1950s Bernard Kettlewell, based at Oxford, demonstrated that the camouflage was adaptive for *typica* and *carbonaria* against a variety of bird predators, firstly by placing reared moths of both phenotypes on tree trunks at both a 'clean' and a polluted site and observing predation, secondly by estimating survivorship of both forms at both sites by releasing and recapturing marked moths, and thirdly by estimating frequencies of *typica*, *insularia* and *carbonaria* in wild population samples. In clean areas *carbonaria* is usually absent, *insularia* can be regarded as a 'suburban' form, and *carbonaria* predominates in polluted towns and cities. Subsequently a team from Liverpool University have shown reduction of the frequency of *carbonaria* in their area with the introduction of clean air legislation.

There the textbook story often ends, but Kettlewell also attempted to show that dominance of the *carbonaria* gene had evolved under selection, much as Clarke and Sheppard had done with butterflies. He outcrossed heterozygous *carbonaria* with *typica* from Canadian moths from an area where there was no *carbonaria* and got immediate breakdown of dominance. The heterozygotes were intermediate and highly variable. But recent similar crosses with mainland European populations have not corroborated his results. The other thing that may have evolved is the *expression* of the *carbonaria* gene. Specimens from Victorian collections show traces of the white patterning of *typica*. We do not know whether they are heterozygous, in which case there are further evidence of the evolution of dominance, or homozygous in which case there are evidence of increase in intensity of expression of the gene.

Darwinism triumphant?

Thus if Kettlewell's interpretation of his results is correct, we have an example where natural selection has produced (1) very rapid (and reversible) changes in frequency of the morphs of the peppered moth, (2) the evolution of dominance, and, possibly (3) evolution of the expression of a gene. Similarly it is difficult to explain the existence of butterfly mimicry by any mechanism other than selection and in the case of mimetic polymorphism, evolution of dominance has also pretty certainly evolved by selection. Many other cases of probable selection have been investigated since the Second World War, some, such as insecticide resistence in insect plant pests, antibiotic resistence in bacteria and resistence to soil pollution by metallic poisons by some grasses, due to human factors. But demonstration of selection is difficult to distinguish from interpretation as selection, and there was a danger of the Synthetic Theory hardening into dogma. It was fortunate, however, if one regards constructive scepticism as an essential ingredient in science (as I do), that a series of challenges to the theory arose successively in the 1960s, the 1970s, and the 1980s. I will describe these in chapter 12.

notes

Dobzhansky's *Genetics and the Origin of Species* and **Mayr's**

Systematics and the Origin of Species were recently (1982) reprinted in their first editions by Columbia University Press (their original publishers). Both went through several editions and finally complete revision and a name change (T. Dobzhansky, *Genetics of the Evolutionary Process*, Columbia University Press, 1970, and E. Mayr, *Animal Species and Evolution*, Harvard University Press, 1963). **G.G. Simpson**'s *Tempo and Mode in Evolution* (Columbia University Press, 1944) was updated and expanded as *The Major Features of Evolution* (Columbia University Press, 1953) but lost its original thrust and cogency in the process. **Julian Huxley**'s *Evolution the Modern Synthesis* had a less radical second edition (Allen & Unwin, 1963). The 'hardening' of all these founding fathers of the Synthetic Theory into panselectionist orthodoxy is discussed by Gould in Grene (1983 – see notes to ch. 8) and in Mayr and Provine (eds), *The Evolutionary Synthesis* (Harvard University Press, 1980) a collective historical work.

Recent work on speciation is described in **D. Otte** and **J.A. Endler** (eds), *Speciation and its Consequences* (Sinauer, 1989). The classic work on mimicry, polymorphism in *Cepaea*, and industrial melanism is described in **Ford**'s *Ecological Genetics* (notes to ch. 10) but recently the *Biological Journal of the Linnean Society* has devoted numbers to the last two (vol. 14, nos 3-4, 1980 and vol. 39, no. 4, 1990, respectively).

Allopatric: two populations (usually of the same or closely-related species) that have geographically-separated ranges. If the ranges are distinct, but have a common boundary with no physical barrier, they are *Parapatric*; if the ranges overlap partially or completely, the populations are *Sympatric*.

(Genetic) Polymorphism: '...the occurrence together in the same habitat at the same time of two or more forms of a species in such proportions that the rarest of them cannot be maintained merely by recurrent mutation'. [E.B. Ford]

12: challenges to the Synthesis

> For many years population genetics was an immensely rich and powerful theory with virtually no suitable facts on which to operate. It was like a complex and exquisite machine, designed to process a raw material that no one had succeeded in mining.... Quite suddenly the situation has changed. The mother-lode has been tapped and facts in profusion have been poured into the hopper of this theory machine. And from the other end has issued – nothing.
>
> (R.C. Lewontin, *The Genetic Basis of Evolutionary Change*, 1974)

the 'balance theory'

At the beginning of chapter 10 I drew attention to the very different view of the nature of the genome taken by Muller and Dobzhansky. According to Muller the ideal genome would be homozygous for the 'best' allele at almost every locus, the 'classical theory' of the genome. In contrast to this was the 'balance theory', as proposed and also named by Dobzhansky in which there was considerable variation in wild populations, which would allow selection to act creatively in the event of environmental change. At first Dobzhansky thought that in a stable environment much of the variation, including that represented by chromosome polymorphism (pp. 120-1) was selectively neutral, that natural selection could not 'choose' between different morphs. But in the 1940s influenced by the developing Synthesis, by laboratory experiments of rearing rival morphs under different temperature regimes, and by the mathematical calculations done for him by Sewall Wright in the GNP series of papers, he began to believe that polymorphism in *Drosophila pseudoobscura* was maintained by selection. There were several possibilities; one was the type of environmental variation within the range of a single population that we saw in Cain and Sheppard's work on the banded snails *Cepaea* (pp. 143-4), which could be in time (i.e. through the year) as well as in space. Dobzhansky and his colleagues showed that the varying success of two chromosome morphs relative to ambient temperature and crowding might explain their both remaining in the population. Another possibility was frequency–dependent selection. The idea here is that in any given environment there is an equilibrium frequency for two alternate morphs – if they were alternate alleles it could be [say] **p** = 70%, **q** = 30%. If either frequency rises significantly above the equilibrium level, the other is favoured by selection; thus the system returns to equilibrium. But by far the most popular explanation was heterozygote advantage (p. 119).

genetic fitness

In the 1950s, however, population geneticists began to worry about the effect on the fitness of whole populations if numerous different chromosomal and genic polymorphisms were retained by heterozygote advantage. Their mode of worrying was to take Muller's metaphor of 'genetic load' and to give it a mathematical formulation. To explain this in the simplest possible terms, we must return briefly to population genetics and the Hardy–Weinberg equilibrium. At equilibrium, and thus with no selection, it will be recalled (pp. 123-4) that the genotype frequencies for two alleles at one locus are:

$$\frac{AA}{p^2} \quad \frac{Aa}{2pq} \quad \frac{aa}{q^2}$$

In modelling selection the 'best' genotype(s) is/are accorded a fitness of one: the genotype(s) which are less fit vary in fitness between zero and one, expressed as 1-*s*, where *s* is the 'selection coefficient'. Thus for heterozygote advantage the fitnesses are:

$$\frac{AA}{1-s_1} \quad \frac{Aa}{1} \quad \frac{aa}{1-s_2}$$

But having homozygotes less fit than the heterozygote will diminish the fitness of the whole population. If the less fit genotypes were eliminated by selection, that would be self-correcting, but with heterozygote advantage they can never be eliminated even if lethal, because in each generation pairing between heterozygotes will yield both homozygotes in the offspring, according to Mendel's laws (pp. 103-6). So heterozygote advantage imposes a segregational (genetic) load on the population. This again can be calculated easily by multiplying fitnesses of the genotypes with their equilibrium frequencies. The result is the net population fitness ($\overline{W}$) for that locus:

$$\overline{W} = \frac{AA}{(1 - s_1)\,p^2} + \frac{Aa}{2pq} + \frac{aa}{(1 - s_2)\,q^2}$$

Thus when $p^2 + 2pq + q^2 = 1$, $\overline{W} = 1 - (s_1 p + s_2 q)$. But that represents the load for only one locus. To represent the load for the whole genome the fitnesses of all the segregating loci in the population must be multiplied together. Later evidence (see below) suggested that there might be 3,000 such loci in a population. Lewontin (author of our epigraph) calculated that $\overline{W}$ for the aggregate would then be about 10^{-43} where the average coefficient was 10%. If that were true an individual with the best genotype at every locus would produce 10^{43} eggs!

neutralism

There are, as Lewontin fully realised, all sorts of invalid simplifying assumptions in such a calculation, notably that there is no epistatic, or any other sort, of interaction between loci. But in the late 1950s a number of population geneticists, notably J.F. Crow in the USA and Motoo Kimura in Japan, began to suggest that a lot of the heterozygosity in natural populations was selectively neutral: there was no fitness differences between the genotypes segregating at any given locus. If this applied at every locus it would eliminate the problem of genetic load – *but at the price of eliminating the Theory of Natural Selection!* Unless, of course, selection acted only on genetic recombination and chromosomal mutations. Nobody, certainly not Crow or Kimura, believed that all the obvious adaptive features of organisms, from the wing pattern of mimetic butterflies to the fish-like shape of a whale, or the succulent, water-retaining leaves of a cactus to the parachute-like seeds of a dandelion, had arisen by random genetic drift. But it was suggested that much or all of the enzyme polymorphism in wild populations was selectively neutral. Many enzymes, the catalysts of all the body's biochemical reactions, occur in two or more versions. The active site of the molecules will be the same, but the 'tail(s)', long strings of amino acids, the building blocks of protein, will have one or more amino acid different in each version of the enzyme. Different versions of the same enzyme came to be known as '*isozymes*' and if both (or all) were coded for at the same gene locus, as '*allozymes*' (from 'allele' and 'enzyme').

The reason that enzyme polymorphism was picked on was that the prevalent view of gene action was that every gene active in promoting chemical synthesis coded for an enzyme – *the one-gene-one-enzyme* theory. But before the mid-1960s nobody had any real idea of the extent of enzyme polymorphism in natural populations, or how to estimate it. And then, suddenly they did – in Lewontin's metaphor 'the mother-lode [had] been tapped'.

enzyme polymorphism

Richard Lewontin was one of the people who tapped it. In 1966 he published two papers with his colleague at Chicago, Jack Hubby, in which they investigated enzyme polymorphism in *D. pseudoobscura*. In the same year Harry Harris in London used the same technique on human enzymes. The technique was *gel electrophoresis*. Any enzyme molecule has, of course, a mass which is the sum of the weights (or more correctly masses) of all the atoms that compose it. But it also has an electric charge, and the 'shape' may differ from that of its allozyme(s). These factors, taken together, determine its mobility due to a (direct) electric current through a slab of gel – starch or plastic (such as acrilamide) in colloidal solution in water containing salts to make it conductive. A current is passed through the gel for a fixed period of time. In the case of *Drosophila* the enzymes do not have to be extracted from the flies. Each fly is ground up and placed in a transverse slit in the gel, so that there is a series of specimens in a line across. At the end of the run the enzymes are revealed by using a dye which is a compound of stain and a specific enzyme substrate, the chemical on which the enzyme acts in life. Two alternate allozymes will stain with the same dye but will have travelled different distances along the gel.

From their results Lewontin and Hubby estimated that an average *Drosophila* population is polymorphic at more than 30% of enzyme loci, with as many as six alleles at a locus. Later, in 1976

Lewontin and two colleagues were able to show that this was almost certainly a gross underestimate. By using a variety of tests on one particular locus they showed that it had not six allozymes but 37, with nearly three quarters of the population estimated to be heterozygous at that locus. It was also shown that at the same locus the Bogatá population (p. 138) was polymorphic at the same locus for a series of allozymes that occurred nowhere else – biochemical divergence by genetic drift leading to species differentiation.

In the 1960s and 1970s there was intense debate between selectionists and neutralists, which also fed back into taxonomy. Random mutation, like random tossing of a coin, would yield statistical regularity, in this case the 'molecular clock' which is assumed by those using distance measures in molecular taxonomy (p. 67). Now the argument has reached a stalemate or perhaps a draw. Even the most ardent selectionists admit that allozymes at one locus may be selectively neutral with respect to one another, but would assert that natural selection of adaptive phenotypic characters (which also represent genotypic differences) is the stuff of evolution. Whether it is the stuff of speciation is more doubtful.

selfish genes

At the same time that the selectionist–neutralist controversy was raging another argument about the potency of selection got under way. The protagonists on both sides of the argument were students of animal behaviour. It started as a controversy about the units on which natural selection could be said to act, but the selectionist side of the argument soon became a method for demonstrating that every aspect of animal behaviour, adaptive, apparently non-adaptive, or just plain puzzling, must owe its origin to natural selection. Today it has become an irritating cliché of television natural history programmes that all the behaviour seen in any animal is conditioned by the necessity for that animal to 'pass on its genes'.

As an example, in lions, males will often take over a pride consisting of females and young from which the incumbant male(s) have been ejected. It has been recorded that under such circumstances the arriving male may kill infants which are not its own progeny. In a book derived from the television series, *Trials of Life*, David Attenborough proffers the following explanation:

> The explanation...is that the male lion, like all individual

> animals, is concerned not with the species as a whole but with the propagation of his own particular lineage, his own genes. Cubs fathered by others have no claims on his affections or support. It is only his own that he wishes to perpetuate. Since he is driven to behave in this way by the influence of his genes, it could be said that it is the genes themselves that are working to ensure their own survival.

Let us set aside the egregious way in which the whole is expressed. It is doubtful whether the average male lion conducts his mating behaviour with a concern for 'the propagation of his own particular lineage' let alone has a grasp of the gene theory of heredity. The last sentence of the quotation embodies a view of natural selection which sees the genes as units of selection in a way made famous, or notorious, by Richard Dawkins in his (1976) *The Selfish Gene.* The idea of genes as the units of selection did not originate with Dawkins: he was preceded by *Adaptation and Natural Selection* (1966) by George C. Williams. Nobody questions the fact that if selection acts at all it must act on individual phenotypes, individual animals and plants. It is they who, in 'the struggle for existence', are healthier, fleeter, longer-lived etc. than their rivals and thus reproductively more successful. But individuals die – it is their descendents who are the beneficiaries.

But according to the selfish gene school of thought genes do not die; they are passed from parent to offspring in the 'germ-plasm', to use Weismann's phrase (p. 100), as well as being replicated in every cell of the individual organism to act as the program for development of all its hereditary features. Thus it is said that if there is a complex of genes for infanticide of non-relatives by (and in) male lions, that will result in the maximum perpetuation of those and all the other individual male's genes by guaranteeing that the pride has a maximum number of his own offspring (by the pride's females) and a minimum number (down to zero) of the offspring of other males. The 'infanticide gene complex', if it exists, is of no benefit to the whole lion species, *Panthera leo*, nor is it of any benefit to the individual pride, in fact quite the reverse, as it tends to reduce genetic variability as well as, possibly, reducing pride numbers, but it will sweep through a population because it produces behaviour that enhances its own frequency.

Here, obviously is the basis of a powerful research programme for those, the 'panselectionists', who wish to attribute every feature of an organism to natural selection. Not only can one attribute obvious adaptations to selection, but anomolous features, often of behaviour in animals, can be explained on the selfish gene principle. The research programme also accords well with population genetics, which models natural selection not in terms of individuals or their phenotypes in populations, but in terms of gene and genotype frequencies.

group selection

One suggested form of selection could not be modelled by population genetics in terms of the spread of the fittest genotype: it was '*group selection*'. In 1962, V.C. Wynne-Edwards, working in Aberdeen, suggested in *Animal Dispersion in Relation to Social Behaviour*, that if a small population of animals occupied a defined area with limited resources it would be of advantage to the population if its numbers could be adjusted so that in any generation the resources were not exhausted with resulting extinction of the whole population. The limiting resource was usually taken to be food, but its limits would be expressed as a maximum number of (bird) nesting sites or a maximum number of individual territories, each of a size to sustain its owner throughout the season. Either every adult member of the population, or, perhaps more commonly, every male would have a rank in a 'peck-order'. If resources were low in any given breeding season or if adult males outnumbered territories, then low-ranking males would fail to occupy a territory and thus fail to breed.

A peck-order certainly exists in many mammals, birds and other animals and the failure to breed could simply be the result of competition, but Wynne-Edwards thought that an accurate matching between population (or group) and resources, adjustable to circumstances, had arisen by competition, i.e. selection, not between individuals of the same species but between groups – self-sacrifice, at least as far as breeding was concerned, was somehow programmed into the genomes of the whole group. Furthermore he believed that intraspecific displays, such as bird song and much eye-catching behaviour had evolved as a means whereby the individual animals could estimate their own population density and perhaps sacrifice their breeding 'rights'.

There is no doubt that such a mechanism, if it existed, would benefit the group. Groups without such inbuilt restraint would at times of famine, or excess population numbers, probably go extinct. But not only is it true that group selection, competition between groups, is difficult to model, it is also difficult to see how it could override individual selection. A mutant 'selfish' male not programmed to group restraint would out-compete all his fellows, even if that resulted in the extinction of the whole group, and there are other problems associated with the theory.

kin selection

Nevertheless Wynne-Edwards' ideas directed the attention of animal behaviourists towards the difficulty of explaining the evolution of 'altruism'. I put that word in quotes because of the danger, which is often the case when words describing human behaviour are used in biological jargon, of assuming that animals, presumably behaving instinctively, are indulging in voluntary action. In group selection theory low-ranking males were behaving 'altruistically' by forebearing to breed, but the bird which gives a warning cry to warn the whole flock of an approaching predator is also displaying 'altruism' as is, in the extreme case, the worker bee who will never breed but spends her life in domestic chores and rearing someone else's (the queen's) children – the reader will see how easily an animal behaviourist can slip into anthropomorphism!

In 1964, however, W.D. Hamilton announced that he had solved the problem of the evolution of 'altruism' by normal natural selection – one could almost hear the community of animal behaviourists heave a collective sigh of relief! Hamilton's concept was that of '*inclusive fitness*'. It concerned degrees of relatedness of individual animals and thus their chance of sharing parts of their respective genomes. Thus a son or a daughter derives half his/her genome from father and half from mother. The 'degree of relatedness' (r) of (e.g.) mother–daughter is a half – there is a 50% chance of the daughter inheriting a particular maternal gene; r for grandmother–granddaughter is a quarter. One can also see that for full sibs (same parents) r = half, and for first cousins, r = one-eighth. Given such measures of relationship and the view that selection is 'passing on ones genes', 'altruism', conferring benefits on another individual to ones own detriment, seemed explicable. Thus the mammal or bird in a flock that gives the warning cry and thus exposes itself to risk, may be saving the lives of related creatures with genes for the same behaviour. This is formalised by 'Hamilton's Rule': $rb-c>0$. r is the degree of relatedness (as above) of the beneficiary of any altruistic act; b is the benefit to the beneficiary (or 'recipient') and c is the cost to the actor. b and c are modelled in terms of reproductive success. If $rb-c$ exceeds zero then selection will perpetuate the 'altruistic' behaviour.

The concept of inclusive fitness is derived from Hamilton's Rule. It is modelled in terms of alleles at a particular locus and thus of a particular genotype:

$$W_i = a_i + \Sigma_j r_{ij} b_{ij}$$

where W_i is the inclusive fitness effect of the genotype i, a_i is the direct effect of the trait on the actor (the chance of being killed by giving a warning cry) ignoring any reciprocal action from others, r_{ij} is the degree of relatedness between the actor and any given member of the flock and b_{ij} the benefit (which could be negative) to that member. In non-mathematical terms the inclusive fitness of a trait is the reproductive potential of an individual with that trait in a given environment (in the absence of help or hindrance from conspecifics) plus the sum of the increase or decrease in reproductive potential produced by the actor on other members of the population multiplied by their respective relatedness to the actor. Thus it was said that a reduction of fitness of the 'altruistic' actor could be more than counterbalanced by the benefits accruing to its relatives. If that were the case the 'altruistic genotype' would be selected for. This was *kin selection* and was regarded as a satisfactory alternative theory to Wynne-Edwards' group selection, because selection was now taken to be on individual organisms rather than groups.

But supposing the altruistic actor displayed its 'altruism' towards non-relatives? The theory of 'reciprocal altruism' was then drummed up by Robert L. Trivers – selection for an instinctive 'I'll scratch your back if you scratch mine' genotype.

social insects

One of Hamilton's greatest triumphs was said to be an explanation in terms of kin selection of the evolution of sterile workers in social insects such as honey bees (see above). Worker bees are all female. In the genome of the female bee there is a switch mechanism such that if she is fed on 'royal jelly' throughout larval life, as an adult she will be a reproductive queen, but if switched to 'bee bread' (a mixture of pollen and nectar) at an early stage, she will be a sterile worker. Ironically 'royal jelly' is secreted by adult workers who also do all the feeding. But how could sterility possibly evolve? Hamilton related it to the peculiar sex-determination mechanism of the honey bee. Male bees (drones) are haploid, that is they have only one set of chromosomes derived intact from their mothers – they have no fathers. (Note that this is an asymmetrical relationship, *r* for a drone's relationship to his mother is one, but *r* mother → son is half.) But when drones themselves become fathers, all their genes are passed on to their daughters, whereas those daughters inherit only half their mother's (diploid) complement of genes. For this reason full sisters (sharing the same parents) have an *r* of three-quarters instead of the usual half – they have an identical complement of genes from their father. Sterile animals have a simple genetic fitness of zero; they cannot reproduce. But a worker bee acquires an inclusive fitness if she rears reproductive sisters (potential queens) which is enhanced by the high *r* for sister–sister relationships. Hence, perhaps, the origin of workers in a haplodiploid species like the honey bee, but the story gets more complicated. Not all the workers in the hive are full sisters. The founding queen may mate with many males in her nuptial flight, before settling down to found a colony with the stored sperm for the rest of her reproductive life. It gets even more complicated if other social Hymenoptera (ants, bees, and wasps) are taken into account, and worse still if termites are included. All termites are social, with sterile workers, but the colony has a king and a queen, workers are of both sexes, and termites are normal diploids not haplodiploids.

'the Panglossian paradigm'

The concepts of gene selection and inclusive fitness are useful interpretive tools, but they illustrate a mode of thought that some students of evolution find objectionable – the 'panselectionism' referred to above. The *a priori* assumption is made that every trait in an organism, whether anatomical, physiological, behavioural or whatever is optimally adaptive. Other examples include the concepts of 'optimal foraging', the assumption that, *ceteris paribus*, an animal will feed in such a way as to yield maximum nutritive input for minimum expenditure of effort. Another example is the 'evolutionary stable strategy' (ESS), derived from game theory, which assumes that any inherited behavioural strategy, as in the response to aggression from another member of a population, will be such that no alternative strategy could 'invade' the genome of the population. Thus if Wynne-Edward's group selected 'altruism' were a fact (as above, p. 153), it would not be an ESS because a 'selfish' individual, who was genetically programmed to occupy a territory and reproduce whatever the resources of the whole population, would be selected.

The optimisation approach then is to assume that adaptation is optimal unless proved otherwise and if proved otherwise to search for extrinsic factors which reduce optimality. Thus it may be suggested that a flock of birds are not foraging optimally because they must be on the alert for predators. An attempt is made to quantify this and the model is adjusted accordingly. But the *a priori* assumption of conditional optima adaptation is not questioned. In a notorious paper criticising this panselectionist approach ('the Panglossian paradigm') Gould and Lewontin summarise their objections thus:

> An adaptationist programme...is based on faith in the power of natural selection as an optimizing agent. It proceeds by breaking an organism into unitary 'traits' and proposing an adaptive story for each considered separately. Trade-offs among competing selective demands exert the only brake upon perfection; non-optimality is thereby rendered as a result of adaptation as well. We criticise this approach and attempt to reassert a competing notion...that organisms must be analysed as integrated wholes [and] that the constraints [of heritage, development and architecture] become...more important in delimiting pathways of change than the selective forces.... We fault the adaptationist programme for its failure to distinguish current utility from reasons for origin...; for its unwillingness to consider alternatives to adaptive stories; for its reliance upon plausibility alone as a criterion for accepting speculative tales....

Thus history, physical constraints and modes of development must be taken into account in studying the nature of organisms.

But another, perhaps more important criticism of the theory of Natural Selection is that it has nothing to say about the pattern of evolution, i.e. phylogeny. I will consider it in the remaining two chapters.

notes

The impact of the discovery of rich enzyme polymorphism on the 'classical theory' of the genome and ideas of genetic load is discussed in **Lewontin**'s *The Genetic Basis of Evolutionary Change* (Columbia University Press, 1974). The neutral theory is described by **Kimura** in a *Scientific American* article (vol. 241 [May 1979] 95-104) and in his book *The Neutral Theory of Molecular Evolution* (Cambridge University Press, 1983). Genes as units of selection appear in **G.C. Williams'** *Adaptation and Natural Selection* (Princeton University Press, 1966) and become selfish in **Richard Dawkins'** *The Selfish Gene* (Oxford University Press, 1976 – 2nd edn, 1989) and **David Attenborough**, *Trials of Life* (BBC Publications, 1990).

Group selection was proposed in **V.C. Wynne-Edwards'** *Animal Dispersion in Relation to Social Behaviour* (Oliver and Boyd, 1962) and he returned to the theme in *Evolution by Group Selection* (Blackwell, 1986). The orthodox, kin selection, explanation, together with other selectionist approaches to behaviour are set out in **J.R. Krebs** and **N.B. Davies** (eds), *Behavioural Ecology – an Evolutionary Approach* (3rd edn, Blackwell Scientific Publications, 1991) to which **S.J. Gould** and **R.C. Lewontin**, 'The Spandrels of San Marco and the Panglossian paradigm: a critique of the adaptationist programme' (*Proceedings of the Royal Society* B205: 1979, 581-98) is an excellent counterbalance.

Isozyme (Isoenzyme): different molecular forms of what is functionally the same enzyme. If two or more isozymes are coded for by different alleles at the same locus, they are *Allozymes*.

13: patterns of evolution

the wrong theory?

> Darwinism, neo-Darwinism, punctuated equilibrium, epigenetic neo-Darwinism, non-equilibrium thermodynamic theories of evolution, and even the various versions of the creationist theory of life all share another shortcoming as explanations of evolution. Each can equally well explain any evolutionary history with a minimum of empirical constraint. None of them uniquely determines one hierarchy. They prohibit no theory of relationship.... A theory of process that is not deterministically tied to the pattern it seeks to explain is no explanation of that pattern at all, but only an explanation of patterns of that kind...selection theory, for example, will explain an incorrectly devised phylogenetic statement with the same facility that it will a correct one because it can explain all statements of that kind.
>
> (Donn E. Rosen, 1984)

Charles Darwin died in 1882, so, inevitably in 1982 there was a flurry of necrophiliac celebrations of the centenary of his death of the same sort we have seen more recently to mark the Mozart bicentenary. Most of the meetings and books commissioned for the Darwin centenary were historical or were collections of essays from those working within the paradigm of the Synthetic Theory, but some publications, including that in which the source of our quotation from Donn Rosen appeared, were for writings by the sceptical, from constructive doubters to the lunatic fringe.

Rosen, until his untimely death a few years ago, was a distinguished ichthyologist (fish systematist) and in the van of the development of 'transformed cladistics'. His opinions bring us full circle back to the *explanans/explanandum* problem to which so much of the early part of this book was devoted. The theory of evolution was proposed to explain the phenomenon and the pattern of natural classification. Should not the theory of evolutionary mechanism then explain, or at least in some way constrain, the pattern of phylogeny, the *explanans*? Natural Selection is a successful, and pretty certainly true, explanation of adaptation, and together with mutation of all sorts, genetic drift, and the neutral fixation of alleles, probably provides and adequate explanation of anagenetic evolutionary change. The role of selection in speciation is less certain and our understanding of the phenomenon less secure. When we come to the role of selection in the explanation of the one true phylogeny, the unique pattern of the development of the rich diversity of life on earth, then the synthetic theory is silent.

Orthodox responses to this sad fact are various. One curious but common one, much favoured by population geneticists and optimisation enthusiasts, as well as numerous philosophers of science, is that evolution is nothing but changes in gene frequency. Rosen quotes the first edition of probably the best undergraduate text on evolution:

> ...biological evolution can be defined...as any change from one generation to the next in the proportion of different genes.
> (Futuyma, *Evolutionary Biology*, 1979)

he also quotes the population biologist, Michod, writing in 1981:

> '...gene changes are evolution' captures in essence the population genetics approach as well as providing the basis of a formal framework within which population genetics as well as other evolutionary biologists work.

and, surprisingly, bearing in mind their theory of punctuated equilibria (see below), and Gould's critique of the 'Panglossian paradigm', this from Eldredge and Gould in 1977:

> Evolutionary change is the modification through time of genes and gene frequencies...

to which I can add this from a useful little book on mathematical population biology (of which the senior author is the inventor of 'Sociobiology', an attempt to take over social science by optimisation theory):

> Evolution can be broadly defined as any change in the genetic constitution of a population. Population genetics has allowed a more precise definition: ANY CHANGE IN GENE FREQUENCY.
>
> (Wilson and Bossert, *A Primer of Population Biology*, 1971 [*their capitals*])

modes of evolution

In his contribution to the Synthetic Theory, *Tempo and Mode in Evolution*, the palaeontologist George Gaylord Simpson proposed three 'modes' of evolution. The first two, phyletic evolution (or anagenesis) and speciation (or cladogenesis) should already be familiar to the reader. The third, very much played down in the (1953) revised version of his book, he called 'Quantum Evolution'. It was 'applied to the relatively rapid shift of a biotic population in disequilibrium to an equilibrium distinctly unlike an ancestral condition'. A population within a species which had been forced into disequilibrium with its environment, presumably because of environmental change or dispersal to a new environment, would either go extinct or evolve rapidly into some new adaptation. Simpson developed the metaphor of adaptive zones, not necessarily topographical or geographical, but representing different suites of adaptations. The disequilibrium then represented an unstable phase between one adaptive zone and another. A large scale example, later suggested by Simpson, was the origin of the ancestral penguins. Penguins have short stubby wings incapable of aerial flight, but they 'fly' powerfully under water. A possible ancestral group is the diving petrels, which can fly both in the air and under water. The quantum evolutionary 'leap' would be the result of natural selection for submarine flight overcoming stabilising selection for both types of flight and, subsequently, perfection of the submarine adaptation. In 1944 Simpson believed that such quantum shifts in adaptation could be at various scales, from within species to those that give rise to major taxonomic groups. Quantum evolution was taken to involve a relatively small number of individuals and, in geological terms, to be very rapid.

punctuated equilibria

In 1972 two young palaeontologists, Niles Eldredge and Stephen J. Gould, suggested that at the level of speciation something like Simpson's quantum evolution was the norm. Simpson's theory was proposed to explain the origin of major new body plans, Eldredge and Gould's 'punctuated equilibrium', to explain what they saw as the usual pattern of the fossil record. To the palaeontologist interested in evolution the ideal fossil record of an evolving group of [say] animals had been taken to be an ancestor–descendent series showing gradual anagenetic change through a sequence of rock strata representing millions of years. In fact Simpson amongst others discussed how with such gradual change one could decide when one species had ended and another begun – a continuum was to be divided into a series of 'chronospecies'. But stratigraphic palaeontologists, those mainly interested in the use of fossils to date rocks, knew that the fossil record was rarely if ever like that. A fossil species appears suddenly in the record, often at the beginning of a recognised fossil time zone, persists for sometimes many millions of years, then is replaced by, or co-exists with, a new closely-related species. But there is no recorded transition between the two and no directional anagenetic change in either.

These uncomfortable facts were routinely blamed on the lamentable incompleteness of the fossil record, a complaint initiated by Darwin. But Eldredge and Gould suggested that the record should be taken at face value. The evolutionary norm was for species to show *morphological* stasis (there is seldom evidence for or against change in physiology or behaviour – or genome) punctuated by change, occurring too rapidly and in too small numbers of organisms to be preserved, at speciation. Their model of speciation was Mayr's allopatric speciation (p. 137) with the founder principle. A small peripheral isolate from the main species stock would differentiate rapidly and thus not reappear in the fossil record until it had reached large numbers and its own morphological stasis.

In 1977 Gould and Eldredge reviewed cases of supposed 'phyletic gradualism', including several standard examples taught to students for decades, such as supposed change in the oyster *Gryphaea* from the Jurassic Period and in the heart urchin *Micraster* from the English Chalk (Cretaceous Period) and found them unsatisfactory or downright false. Ironically the only case they could find of phyletic gradualism which met their strict criteria was that of a fossil foramniferan from the Permian of Japan studied by T. Ozawa. He showed a steady increase in size of the shell entrance in these minute shelled Protozoa over a period of some 10 million years. But the irony was that these are asexual organisms – speciation as a result of reproductive isolation does not occur.

species selection

The proposal of punctuated equilibrium caused a tremendous furore, which was enhanced when another paleaontologist, Steven Stanley, suggested in 1975 that the pattern of phylogeny and the nature of trends seen in the fossil record required a theory other than natural selection for their explanation. Stanley accepted the punctuated equilibrium model but went on to suggest that speciation, when most adaptive change took place, was essentially random with respect to the direction of large scale evolution. Accepting the pattern of allopatric speciation of peripheral isolates, the direction of change as the isolate becomes a new species is as much determined by its unrepresentative gene pool and its chance arrival in a new environment, as by anything that would determine the direction of evolution of the larger group to which the new species belongs. In Stanley's theory of '*species selection*', therefore, he likens the randomness of speciation with respect to the direction of macroevolution to the randomness of mutation with respect to the direction of adaptive change within species. The latter is due to differential survival and reproduction of individuals; the former to differential survival and reproduction of whole species. Thus long-lived (and morphologically static!) species would produce more daughter species, particularly if they represented what came to be known as a 'speciose clade'. Any trends in evolution seen in a major group in the fossil record would be due to differential longevity and speciation rate rather than the cumulative selection of adaptations.

Before the proposal of species selection orthodox evolutionists, the population geneticists and the panselectionists, could claim that punctuated equilibria was quite compatible with the Synthetic Theory. Stasis was a mystery, but if most evolutionary change took place during the process of speciation, the change was nevertheless adaptive and produced by the action of natural selection on random mutation and recombination (with no doubt a minor component of genetic drift and the neutral fixation of alleles). Microevolution and macroevolution were 'decoupled in the important sense that macroevolutionary patterns cannot be deduced from microevolutionary principles' (to quote two distinguished selectionists, G. Ledyard Stebbins and Francisco Ayala writing in 1981), but the *mechanism* of evolution could be extrapolated to all levels. If, however, some sort of 'species sorting' (Gould's 1985 phrase) is responsible for the pattern of phylogeny, then the decoupling is total.

extinction

Gould's 1985 paper defines three 'tiers' in evolution, first, 'evolutionary events of the ecological moment' mediated by natural selection etc., second, 'evolutionary trends within lineages and clades that occur during millions of years in "normal" geological time between events of mass extinction', perhaps mediated by 'species selection', and, third, mass extinction. For those who like their history chopped up into decades, one can say (roughly) that the Synthetic Theory was established in the 1940s, hardened into rather smug orthodoxy in the 1950s was challenged by the neutralists in the 1960s, by punctuated equilibria and species selection in the 1970s. Neither of these successive challenges has been resolved, except perhaps that neutralism and selectionism have reached a truce. The controversy of the 1980s, also unresolved, has been mass extinction.

Geologists have accepted for many decades that there have been a number of episodes of mass extinction in the history of life. The most notorious is that at the end of the Cretaceous Period, about 65 million years ago, notorious as marking the end of the dinosaurs, but by far the biggest, i.e. the most catastrophic, was that at the end of the Permian Period, some 250 million years ago. It is estimated that at the end of the Cretaceous rather less than a quarter of the world's species became extinct, whereas the end-Permian 'event' accounted for as much as 95% of all species then living! One major event is recognised between those of the Permian and Cretaceous (end Triassic: ca. 220 million years) and from the base of the Cambrian (570 million years), the time when fossil animals and plants became common in the fossil record, until the end of the Permian, two more.

Two interlinked controversies are still debated. First, were major extinction events due to terrestrial or extra-terrestrial causes? and, second, has there been a periodicity in extinction events with the known major events interspersed with regular minor events? It would be beyond the scope of this book to give more than the merest outline of these controversies, but the reader should note that a regular cycling of events would, if established, suggest some regular extraterrestrial cause. The latter was proposed by father and son, Luis and Walter Alvarez, and colleagues in 1980. Walter Alvarez discovered unusually high levels of the rare element iridium in clays at the Cretaceous/Tertiary ('K/T') boundary in Italy (and subsequently in Denmark and New Zealand) and suggested with his colleagues that this recorded the impact of an enormous meteorite, which would have resulted in something like the then popular scenario of a 'nuclear winter'. Other evidence was added later, of quartz grains showing rapid impact and possible soot from global fires *inter alia* – all suggestive but not conclusive. The 'terrestrial opposition' countered with normal geological processes plus the Deccan Traps, some million cubic kilometres of basalt rock in western India, which had welled out through the earth's crust from the mantle below in K/T times. This too would yield many features of the nuclear winter.

periodicity

The periodicity was proposed by David Raup and Jack Sepkoski of Chicago University on the basis of an enormous data base, compiled from the literature, of the geological dates of first and last appearance of marine animal family-level taxa. They claimed a significant peak in rates of extinction every 26 million years after the Permian event. There have been many statistical and other critiques of their theory, but what I regard as the *coup-de-grace* was administered by Colin Patterson and Andrew Smith of the London Natural History Museum. Patterson and Smith looked respectively at the fish and echinoderm (starfish, sea urchins *et al.*) data, comprising about 20% of the whole data base. For those two groups only 25% of Raup and Sepkoski's families were natural taxa. Compute from those and there was no trace of periodicity. Compute from the phony taxa and the appearance of periodicity was enhanced. But, as always in science, the debate continues.

notes

Donn Rosen's paper 'Hierarchies and history' is in *Evolutionary Theory: Paths into the Future* (ed. J.W. Pollard, Wiley, 1984). For Simpson's two books see notes to ch. 11. Apart from their odd concept of evolution, **Edward O. Wilson** and **William H. Bossert**, *A Primer of Population Biology* (Sinauer, 1971) is a useful introduction to population genetics (*inter alia*). The fully developed theory of punctuated equilibria is in **S.J. Gould** and **N. Eldredge**, *Paleobiology* 3 (1977) 115-51. **Steven Stanley** introduces, *inter alia*, his ideas on 'species selection' in *The New Evolutionary Timetable...* (Basic Books, 1981) and an orthodox account of extinctions in *Extinction* (Freeman – Scientific American Library, 1987). In the October 1990 number of *Scientific American* 263 (no. 4, 42-60) there are opposing articles advocating extraterrestrial and terrestrial causes of mass extinction by **W. Alvarez** and **F. Asaro** and **V.E. Courtillot**, respectively.

Other references: **S.J. Gould**, *Paleobiology* 11 (1985) 2-12. **D. Raup** and **J.J. Sepkoski**, *Proceeding of the National Academy of Sciences, USA* 81 (1984) 801-05. **C. Patterson** and **A.B. Smith**, *Nature* 330 (1987) 248-51.

14: questions answered

In this final chapter I will attempt to summarise the content of this book by posing a series of questions about evolution and then to demonstrate the extent to which those questions are answerable and answered.

The first question concerns the *explanandum* of evolution and the success of the *explanans* as a theory explaining that *explanandum*. I have already suggested at the beginning of ch. 6 (pp. 58-9) that the theory was the only plausible explanation of the phenomenon of Natural Classification. Given the modern understanding of heredity developed in subsequent chapters, we now have a better understanding of the mechanism of evolutionary change and this further corroborates the *explanans*.

My next question (pp. 73ff.) concerns the nature of the evidence for evolution. Despite the almost universal confusion of *explanandum* with evidence, I once again hope that the reader will agree that the evidence from vestigial organs, biogeography and the fossil record (ch. 7) is not only completely convincing, but that it would require the most extraordinary intellectual contortions (a form of exercise at which the 'creationists' are adept) to explain it in any other way.

The remaining questions take us onto more controversial ground. I will phrase the next one as follows: 'Evolutionary theory implies "community of descent" and thus a unique historical pattern. Can one make any significant generalisations about the nature of that pattern?' We have seen in chapter 5 that traditional 'evolutionary' taxonomy and also phylogenetic cladistics each claim to be reconstructing the pattern of phylogeny. As in virtually every other taxonomic method the result is a divergent irregular hierarchy. Particularly in the case of cladistics, this can be corroborated as a Natural Classification by producing classifications of the same suite of organisms from two or more different sets of data (e.g. anatomy and DNA). Consilience among them suggests that they represent the true pattern. The same applied in the case of transformed cladistics, which gets the priority of *explanandum* over *explanans* right, but at the expense of having to accept the divergent hierarchy *a priori*.

In the case of animals, this is reasonable. Hybrid species are probably rare, speciation produces a divergent branching pattern of phylogeny; so it is reasonable to expect that pattern of classification and pattern of phylogeny will be isomorphic, whichever is given priority. With plants, as we saw in chapter 6, there are complications – only if we had some means of getting at the pattern of phylogeny would it tell us about the frequency of hybrid species, and, as I explained in chapter 6, when considering hybrids the *explanandum/explanans* relationship gets hopelessly tangled from a logical point of view. Different complications ensue, if one tackles the classification of bacteria and other largely asexual organisms. There are (or have been) no speciation events, so that the nodes of a dendrogram cannot represent 'splitting'. It is no coincidence that most bacterial taxonomists are pheneticists.

But what of the one unique pattern of phylogeny? Here we can assert (or reject) the empirical observation that punctuated equilibrium prevails in the fossil record, and with acceptance propose that the punctuation events represent speciation (although P.E. is also known in asexual organisms – foramnifera). As is so often the case in biology, the assertion is that P.E. is the *usual* pattern – all or nothing assertions are best left to the physical sciences.

the mechanism

My remaining questions concern the mechanism of evolution. The first is to ask about the possibility of a single comprehensive theory to explain both anagenesis (evolutionary change) and cladogenesis (speciation). Have we got such a theory? or is such a theory even possible? Natural selection does explain *adaptive* anagenesis and in that respect has no valid rivals, although, as we shall see below, others than Donn Rosen believe it is the wrong sort of theory to explain diversity. The orthodox position is then to extend selection to explain cladogenesis, and to assert that no other theory of mechanism (such as 'species selection') is necessary to explain the whole of phylogeny. But cladogenesis almost certainly requires an allopatric phase between two incipient sister species, to be explained by dispersal or vicariance (p. 87).

One can then ask 'To what is the divergence between incipient species (in form, behaviour and as genetic isolation) to be attributed?' Here, as we saw in chapter 11, it is difficult to make general statements. Genetic isolation can occur entirely in allopatry, as in the Bogatá population of *Drosophila pseudoobscura* (p. 138) in which case, whether due to genetic drift (intrinsic factors only) or selection for different environments (intrinsic and extrinsic factors), it certainly was *not* produced by direct selection for genetic isolation. But in other cases, reviewed in chapter 11, various types of direct selection, against hybrids, or for isolating mechanisms or assortative mating, may occur after the incipient species become sympatric again.

But should any causal factors supplementary to those explaining anagenesis and cladogenesis be invoked to explain the historical diversity of organisms? Mass extinction is a possibility, and events like that at the end of the Permian and of the Cretaceous must indeed have had a profound effect on the future history of life, not just on the relative frequency of groups of organisms, but also on the direction of evolution. The extinction events were 'bottlenecks' in time in the same way, but on a much greater scale, that peripheral isolates in speciation represent bottlenecks in space. The founder principle must also have applied on a grand scale.

In the remaining questions we can consider the merits of a particular theory in explaining anagenesis and/or cladogenesis. First we can ask 'does any given theory invoke both intrinsic and extrinsic factors?' If the given theory is the current version of Natural Selection, then the answer is obviously, yes! The intrinsic factors are gene mutations, chromosome mutations (such as inversions: pp. 120-1), recombination as a result of crossing over and sexual reproduction. If these factors are then expressed as differences in the phenotype of an organism from that of its fellows and that difference is adaptively significant, then there will be differential success in survival and reproduction, i.e. selection. Following that question one can ask whether the direction of evolutionary change, anagenesis, is attributable principally to intrinsic or extrinsic factors. Here selection may be contrasted with (unsuccessful) rival theories. There are two logically possible extremes. If the direction of anagenesis is due entirely to internal factors, then one has an orthogenetic theory like that at first proposed by Lamarck (p. 32) and later taken up by 19th- and early 20th-century palaeontologists (p. 100). But the rest of Lamarck's theory represents (almost) the other extreme. Somehow the environment was supposed to produce a response in an animal which modified the development of that animal in an adaptive and heritable way.

Theories raised as rivals to selection at the beginning of the 20th century raise more interesting problems in the light of questions about the relative importance of intrinsic and extrinsic factors. The saltational theories of evolution (p. 106) espoused by Bateson, de Vries and Johannsen (and later Goldschmidt) are comparable in their decoupling of micro and macroevolution to Stanley's species selection. The newly speciated 'hopeful monster' (p. 140) having survived at all, owes its new status to a 'random' reorganisation of the genome undirected by any environmental factors, but the eliminated 'hopeless monsters' must have been culled by a purely negative natural selection. Since then most opponents of Natural Selection have seen it merely as a negative culling force, eliminating the unfit but creating nothing. We have seen that this need not, in fact certainly is not, the case. Selection is not just a judgement on individual mutant genes. New recombinations, new epistatic or additive effects and even selection for closer gene linkage (via inversions) to produce 'supergenes' (like those controlling the wing patterns in the mimetic butterfly *Papilio dardanus* pp. 140-3) are all creative effects of selection.

Two other reasons have been suggested for the claim that selection is merely a culling force. The first reason is that if the 'classical theory' of the genome, as seen in the 1950s by Muller and other laboratory geneticists (pp. 117-18), were true, then selection could only produce uniformity in a stable environment, giving the 'ideal genome'. As we have seen the classical theory is wrong, but the second reason requires more discussion.

We saw at the end of chapter 12 that, in agreement with Gould and Lewontin's 'Panglossian paradigm' paper, natural selection cannot be regarded as omnicompetent, able to produce the perfect adaptation of any given organism. A view at the other extreme would be that constraints on the possible outcomes of selection, due to previous history, modes of development and possible body plans, are so restrictive, that selection can only direct the evolution of organisms into a series of pre-ordained forms. This assumption can be turned into a positive research programme, as proposed by Brian Goodwin of the Open University. Goodwin accepts that selection occurs but holds that it is the wrong sort of theory for the scientific explanation of the diversity of organisms. He pleads (in the same volume as Donn Rosen's paper from which the epigraph to chapter 13 was taken) for a 'generative paradigm'.

an alternative?

He takes as an example the tetrapod limb and its development. There has been an enormous amount of work in recent years on deciphering the program, ultimately of course a genetic one, which controls the development of the legs of tetrapods. Urodeles (newts and salamanders) have been a favourite material, because, not only can one study normal limb development through the embryo and tadpole stage, but some urodeles are able to regenerate lost limbs. Limb skeleton development seems to follow a number of mathematically simple rules, which can also explain the differences between the limbs of different animals. Change in an equation can model the difference between the fore-limb of a horse, with one functional toe, and that of a human with five fingers. Thus the task of comparative anatomy and embryology, according to Goodwin, is not to seek adaptive reasons for differences between organisms, nor yet to seek homologies and reconstruct history, but to discover the underlying laws of form which yield a finite number of possible patterns.

Goodwin's ideas, and his work together with that of others in this school of thought, draws on the methods of a famous biologist who stood outside the developing genetics and evolutionary theory of the early years of this century, D'Arcy Thompson. His *On Growth and Form* was first published in 1917, with a second edition in 1942, and has been reprinted many times since. Thompson showed how the shapes of organisms could be described by simple mathematical equations. The horns of sheep and goats, the shell of the pearly nautilus (a relative of octopus and squid) and the chambered shells of protozoan foraminifera are all described by equations for logarithmic spirals. Stresses on the girder represented by the backbone of a land vertebrate were the same as those on a cantilever bridge, and were resisted in similar ways. Thompson also showed that pairs of closely related whole fish, or mammals' skulls, of very different shapes, could be compared by plotting one on a square grid and then distorting the coordinates of the grid to give the proportions of the other (Fig. 14.1).

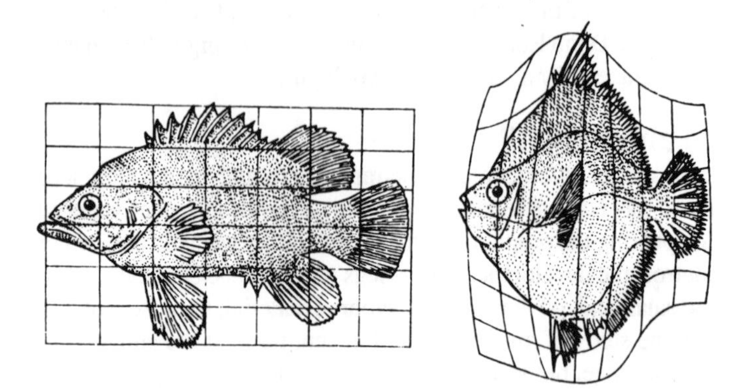

Fig. 14.1 Two acanthopterygian fishes. The proportions of *Antigonia* (right) are represented by distortion of coordinates over *Polyprion* (left). (After D'Arcy Thompson, *On Growth and Form*, Cambridge University Press, 1917.)

the logic of selection

The last three questions I wish to consider, I originally posed at the beginning of a short undergraduate course at Newcastle University on natural selection in order to emphasise the logical consequences of the theory. If, with Mayr, we assume that speciation is probably always allopatric (at least in animals), the questions with their answers are as follows:

1. 'Does selection *always* result in evolution (anagenesis and/or cladogenesis) or does it sometimes result in maintenance of the *status quo*?' Stabilising selection almost certainly maintains genetic polymorphism in cases like that of mimetic polymorphism in butterflies (pp. 140ff.) or the chromosome polymorphism of *Drosophila*, so for anagenesis, the answer to the first part of the question is *no*, and to the second *yes*.

2. 'Is selection *necessary* for evolution...to occur...or can evolution occur without it?' Again, for anagenesis, the answer to the first part of the question is almost certainly, *no*. If allozymes are selectively neutral with respect to one another (pp. 149-51), then their random fixation or elimination could occur and a new mutant neutral allozyme could appear, be retained and spread without reference to selection. Also Wright's shifting balance theory (p. 128 depends in part on features increasing in frequency, at least over short periods, without, or even against, selection.

3. 'Is selection sufficient for evolution...to occur, or do other factors, not proposed in the theory, also have to be invoked?' The answer to this question, for anagenesis, depends on interpretation. If selection is taken to be just the differential survival and reproduction of better adapted and heritable phenotypes, then the answer is no. But if the causal intrinsic factors producing those phenotypes, mutation and recombination in a genetically heterogeneous population, are part of the theory, then the answer is yes!

The answers to these three questions with respect to anagenesis, evolutionary change in taxonomic characters with time, shows the unsatisfactory nature of the theory of Natural Selection as a testable theory. Evolution is explained by selection, but so is the absence of evolution. Evolution can occur by means of selection, but it can also occur independently of selection; and if selection fails to produce evolution in a population, this must be because the hereditary raw material is absent. As Richard Lewontin wrote in 1972:

> ...what good is a theory that is guaranteed by its internal logical structure to agree with all conceivable observations, irrespective of the real structure of the world? If scientists are going to use logically unbeatable theories about the world, they might as well give up natural science and take up religion. Yet is that not exactly the situation with regard to Darwinism?

But then he goes on to say that 'The trouble with this analysis is that even though natural selection might not be an epistemologically satisfactory hypothesis, it might nevertheless be true'.

The theory of Natural Selection can be made rather more epistemologically respectable, if firstly, it is reduced to being an explanation of *adaptive* anagenesis, so that the assertion underlying question (2) is that 'Natural Selection is a necessary cause of *adaptive* anagenesis'. Then, secondly, the proposition can be turned round (as I did in my formulation of the theory: p. 95) to 'Adaptive anagenesis is (always) the result of Natural Selection'. That proposition would be refuted (or at least taxonomically confined – e.g. 'in animals') by an example of the Lamarckian 'inheritance of acquired characters'.

When we consider the answers to my questions 1-3 with respect to cladogenesis, the logical situation is just as unsatisfactory, but oddly our ignorance about speciation signals a more useful research programme. Can speciation occur without an allopatric phase (apart from by polyploidy and hybridisation in plants)? Can full genetic isolation evolve between incipient sister species in allopatry (as has happened in part with Bogatá *Drosophila*)? Does selection for assortative mating (the Wallace effect: p. 138) occur? These and other answerable questions about speciation seem to me to represent a more important research programme than inventing adaptive Just-so stories about behavioural traits.

Darwin's five theories

I noted in a footnote on the first page of the introduction that Ernst Mayr distinguished five theories of evolution rather than two (the theory that evolution has occurred and a theory of mechanism), or, as Darwin himself claimed, a single theory of interdependent parts. Mayr's five, distinguished in an essay in a collective work in 1985, and also in recently published book, *One Long Argument: Charles Darwin and the Genesis of Evolutionary Thought* (1991), are as follows:

1. Evolution as such. This is the theory that the world is not constant nor recently created nor perpetually cycling but rather is steadily changing and that organisms are transformed in time.
2. Common descent. This is the theory that every group of organisms descended from a common ancestor and that all groups of organisms...ultimately go back to a single origin of life on earth.
3. Multiplication of species. This theory explains the origin of the enormous organic diversity. It postulates that species multiply either by splitting...or by 'budding', that is, by the establishment of geographically isolated founder populations that evolve into new species.
4. Gradualism. According to this theory, evolutionary change takes place through the gradual change of populations and not

by the sudden (saltational) production of new individuals that represent a new type.

5. Natural selection. According to this theory, evolutionary change comes about through the abundant production of variation in every generation. The relatively few individuals who survive, owing to a particularly well-adapted combination of inheritable characters, give rise to the next generation.

All evolutionists accept the first theory (by definition!) and the majority of late 19th-century naturalists and biologists were converted to evolution by the *Origin.* But not all evolutionists have accepted common descent. As Mayr points out, Lamarck did not. Lamarck's theory was of continuous events of spontaneous generation with descent from the generated organisms of innumerable parallel evolutionary lines. I know of no 20th-century evolutionist who accepts this view. The biochemical affinity of all organisms – for instance cytochrome c is a variable molecule that can be used for the classification of almost any organism – and the near universality of the genetic code embodied in DNA or RNA suggest common ancestry. A possible exception to that universality is that 'mad cow disease' and other infectious brain disorders may be caused by a purely protein 'organism'.

The third theory, multiplication of species by splitting or budding is opposed by Mayr to belief in the spontaneous origin of species without some division of a population by extrinsic factors. Interpreted this way multiplication would have been rejected by de Vries (p. 107) and Goldschmidt. Today, as we have seen, it is generally accepted for animals – even most sympatric scenarios involve some topographical separation – and, with the exceptions I have noted, for plants.

In Darwin's time gradualism was the most contentious aspect of his views. After the rediscovery of Mendel's work it was rejected by the early geneticists who rejected 'multiplication', de Vries, Johannsen, and of course Goldschmidt. The advocates of punctuated equilibria were accused of being 'saltationists' and seemed at times to be flirting with Goldschmidt-like ideas. But gradualism says nothing about speed of evolution, nor forbids stasis, it simply rejects major morphological jumps. In that sense it is generally accepted today.

Finally we have the theory expressed in a phrase made famous by Darwin (but also proposed without the phrase by Wallace) – 'Natural Selection'. Despite all the objections to it that I have set out in this and previous chapters, and despite the misuse of the concept by those, confused by metaphor, who regard selection as a agent rather than a consequence of extrinsic factors bringing about differential reproduction, natural selection is a theory whose two proposers must be accorded the status of genius.

notes

B.C. Goodwin, 'Changing from an evolutionary to a generative paradigm in biology' is in *Evolutionary Theory: Paths into the Future* (ed. J.W. Pollard, Wiley [1984] 99-120).

D'Arcy Thompson's *On Growth and Form* was first published in 1917 (Cambridge University Press). There is an abridged edition (ed. J.T. Bonner, 1961, pbk 1966). The quotation from Richard Lewontin is from 'Testing the Theory of Natural Selection' (*Nature* 236, 181-2) a review of a Festschrift for E.B. Ford. **Darwin**'s five theories are set out by Ernst Mayr in *One Long Argument: Charles Darwin and the Genesis of Evolutionary Thought* (Harvard University Press, 1991).

Index